T0134846

Anomaly-Detection and Health-Analysis Techniques for Core Router Systems

Shi Jin • Zhaobo Zhang • Krishnendu Chakrabarty
Xinli Gu

Anomaly-Detection and Health-Analysis Techniques for Core Router Systems

Shi Jin
Hardware
Nvidia Corporation
Sunnyvale, CA, USA

Krishnendu Chakrabarty
Electrical and Computer Engineering
Duke University
Durham, NC, USA

Zhaobo Zhang
Futurewei Technologies
Santa Clara, CA, USA

Xinli Gu
Futurewei Technologies
Santa Clara, CA, USA

ISBN 978-3-030-33666-0 ISBN 978-3-030-33664-6 (eBook)
https://doi.org/10.1007/978-3-030-33664-6

This Springer imprint is published by the registered company Springer Nature Switzerland AG.
The registered company address is: Gewerbestrasse 11, 6330 Cham, Switzerland

To my parents,
Dehai Jin and Xiashi Zhu for their endless support.
To my love,
Yicong Mao for her encouragement and unconditional love throughout these years!

—Shi Jin

Preface

A three-layer hierarchy is typically used in modern telecommunication systems in order to achieve high performance and reliability. The three layers, namely core, distribution, and access, perform different roles for service fulfillment. The core layer is also referred to as the network backbone, and it is responsible for the transfer of a large amount of traffic in a reliable and timely manner. The network devices (such as routers) in the core layer are vulnerable to hard-to-detect/hard-to-recover errors. For example, the cards that constitute core router systems and the components that constitute a card can encounter hardware failures. Moreover, connectors between cards and interconnects between different components inside a card are also subject to hard faults. Also, since the performance requirement of network devices in the core layer is approaching Tbps levels, failures caused by subtle interactions between parallel threads or applications have become more frequent. All these different types of faults can cause a core router to become incapacitated, necessitating the design and implementation of fault-tolerant mechanisms in the core layer.

Proactive fault tolerance is a promising solution because it takes preventive action before a failure occurs. The state of the system is monitored in a real-time manner. When anomalies are detected, proactive repair actions such as job migration are executed to avoid errors, thereby maintaining the non-stop utilization of the entire system. The effectiveness of proactive fault-tolerance solutions depends on whether abnormal behaviors of core routers can be accurately pinpointed in a timely manner.

This book first presents an anomaly detector for core router systems using correlation-based time-series analysis. The proposed technique monitors a set of features obtained from a system deployed in the field. Various types of correlations among extracted features are identified. A set of features with minimum redundancy and maximum relevance are then grouped into different categories based on their statistical characteristics. A hybrid approach is developed to analyze various feature categories using a combination of different anomaly detection methods, leading to the detection of realistic anomalies.

Next, this book presents the design of a changepoint-based anomaly detector such that anomaly detection can be adaptive to changes in the statistical features of data streams. The proposed method first detects changepoints from collected time-series data and then utilizes these changepoints to detect anomalies. A clustering method is developed to identify a wide range of normal/abnormal patterns from changepoint windows. Experimental results show that changepoint-based anomaly detector can detect "outliers" even when the statistical properties of the monitored data change significantly with time.

An efficient data-driven anomaly detector is not adequate to obtain a full picture of the health status of monitored core routers. It is also essential to learn how healthy a core router system is and how different task scenarios can affect the system. Therefore, this book presents a symbol-based health status analyzer that first encodes, as a symbol sequence, the long-term complex time series collected from a number of core routers and then utilizes the symbol sequence for health analysis. Symbol-based clustering and classification methods are developed to identify the health status.

In order to accurately identify the health status, historical operation data needs to be fully labeled, which is a challenge in the early stages of monitoring. Therefore, this book presents an iterative self-learning procedure for assessing the health status. This procedure first computes a representative feature matrix to capture different characteristics of time-series data. Hierarchical clustering is then utilized to infer labels for the unlabeled dataset. Finally, a classifier is built and iteratively updated using both labeled and unlabeled datasets. Partially labeled field data collected from a set of commercial core routers are used to experimentally validate the proposed method.

In summary, the book tackles important problems of anomaly detection and health status analysis in complex core router systems. The results emerging from this book provide the first comprehensive set of data-driven resiliency solutions for core router systems. It is anticipated that other high-performance computing systems will also benefit from this framework.

Sunnyvale, CA, USA — Shi Jin
Santa Clara, CA, USA — Zhaobo Zhang
Durham, NC, USA — Krishnendu Chakrabarty
Santa Clara, CA, USA — Xinli Gu

Acknowledgments

Shi Jin and Krishnendu Chakrabarty acknowledge the research grant to Duke University from Futurewei Technologies.

Shi Jin acknowledges Professor Krishnendu Chakrabarty for his constant support and patient guidance. Shi Jin also thanks all the support received from his labmates at Duke University, including Fangming Ye, Kai Hu, Ran Wang, Zipeng Li, Mohamed Ibrahim, Abhishek Koneru, Rana Elnaggar, Zhanwei Zhong, and Mengyun Liu.

Contents

Chapter 1
Introduction

Internet Protocol (IP) networks have developed dramatically during the last decade and have become indispensable in the modern information society. Massive amounts of data are being continuously transferred across the worldwide IP network. Figure 1.1 shows an example of a modern IP network. We can see that data packets are forwarded between computer networks through a set of networking devices called routers. Relentless advancement in hardware and software techniques has increased the performance and complexity of routers by orders of magnitude. With increasing complexity and higher speed, a wider range of faults in routers are becoming more difficult to detect, diagnose and repair in time.

In this chapter, we introduce basic concepts and background related to core routers in modern telecommunication systems. We also discuss prior work on anomaly detection and provide motivation for the book. Sections 1.1 and 1.2 present an overview of network hierarchy and the underlying architecture and working principle of core router systems. Section 1.3 discusses various existing anomaly detection techniques, including an enumeration of their advantages and disadvantages. Section 1.4 describes motivation for the book. Finally, an outline of this book is provided in Sect. 1.5.

1.1 Network Hierarchy

Communication systems are composed of networks and network devices. As shown in Fig. 1.2a, early networks were deployed in a flat topology. In this topology, each network device performs the same function. Therefore, it is easy to design and implement this type of a network. However, as the size of networks increases, the response time increases significantly, severely impairing the quality of services provided by the network [2]. Moreover, modifying or updating a small portion of the network may require replacing a large number of devices, significantly increasing

S. Jin et al., *Anomaly-Detection and Health-Analysis Techniques for Core Router Systems*, https://doi.org/10.1007/978-3-030-33664-6_1

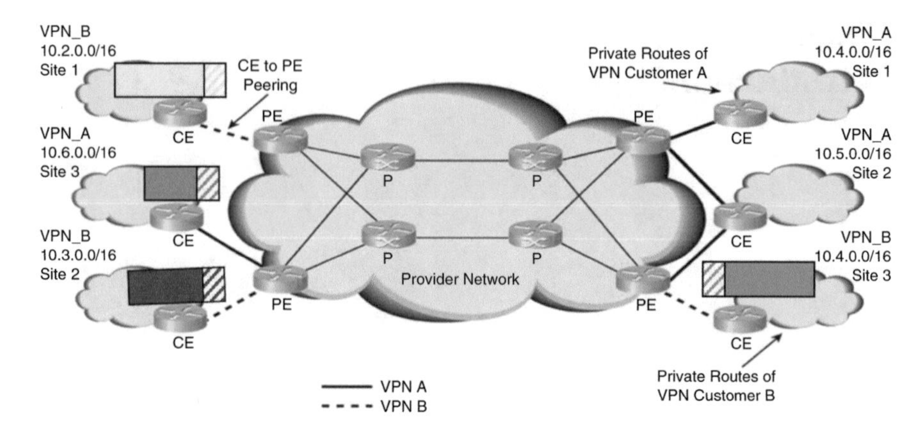

Fig. 1.1 An example of a modern IP network [1]

the maintenance cost. Therefore, a hierarchical network design is needed to ensure high performance and reliability.

As shown in Fig. 1.2b, a hierarchical network design divides the network into three layers, including Core, Distribution (or Aggregation), and Access [2]. Each layer provides specific functions and features within the overall network. The Core layer can be viewed as the backbone—it provides high speed and redundancy. The Distribution layer can be considered as the intermediate boundary—it provides policy-based connectivity. The Access layer can be seen as the entry point—it provides end users access to the network [2]. While this is a simplified view of the network, it provides a general high-level overview that can serve as the basis for reliability analysis and resilience assessment. The benefits associated with such a three-layer hierarchical network designs are as follows [3, 4]:

- Scalability: The modularity of hierarchical design allows the replication of elements in each layer. Expanding the network become easier since devices in each layer have similar roles and functionalities. For example, if the distribution layer contains two devices and each device can support ten devices in the access layer, a total of twenty access-layer devices can be directly added to the network without making any changes to the high-level deployment.
- Redundancy: Redundancy in hierarchical networks is implemented via replication of network elements. For example, path redundancy is achieved by connecting one access-layer device to two different distribution-layer devices. Whenever the running distribution-layer device fails, the access-layer device can switch to the other distribution-layer device, and the entire network will continue to work as usual.
- Performance: High speed of hierarchical networks is achieved by forwarding large amount of traffic in the core layer, avoiding frequent low-speed data transmission in the access layer. The distribution layer first aggregates data received from the access layer, and then transmits them to the core layer after

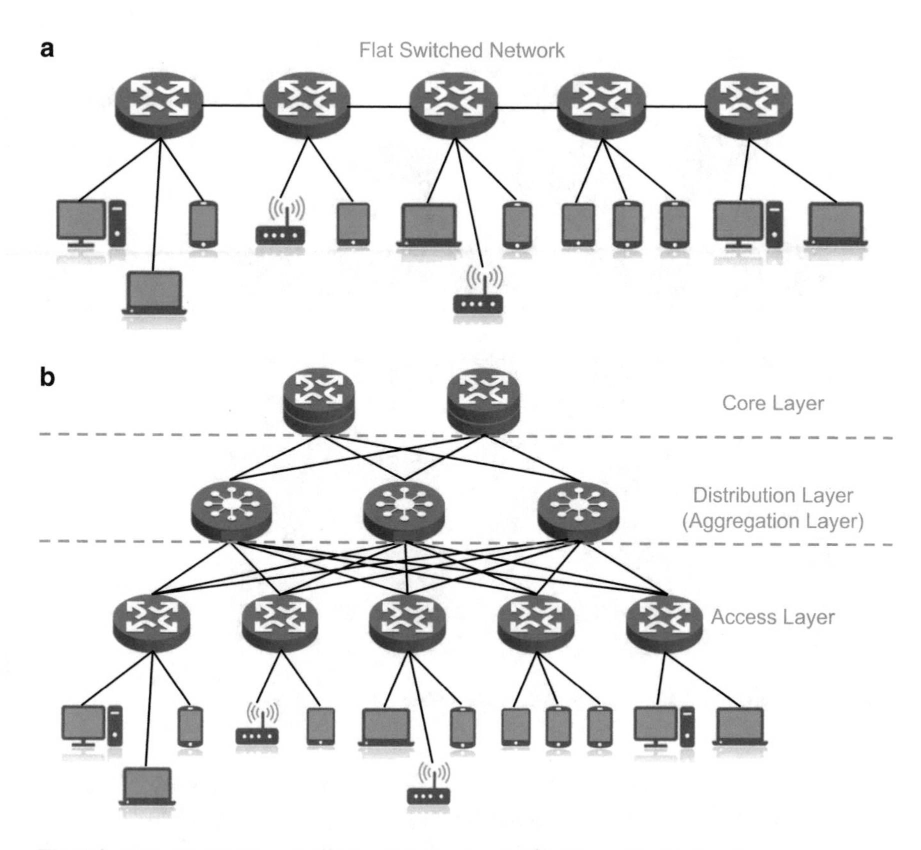

Fig. 1.2 Network topology. (**a**) Flat switched network. (**b**) Hierarchical network

policy-based filtering. Such division of labor greatly reduces the probability of network contention and congestion, enabling near-wire speed between all devices.

- Security: Security is enhanced in hierarchical networks because different security policies can be applied to different layers. For example, access-layer switches can be configured with various port security options that provide control over which devices are allowed to connect to the network. In addition, access-control policies that define which communication protocols are deployed and where they are permitted to go can be applied to the distribution layer to restrict traffic based on higher-layer protocols.
- Manageability: Manageability is improved in hierarchical networks because the functions performed by each layer are consistent throughout that layer. Therefore, any changes in a layer's functionalities can be repeated across all devices in that layer. Also, configurations can be copied between devices in the same layer with minor modifications. Operation logs can be compared between devices in the same layer for simplified troubleshooting.

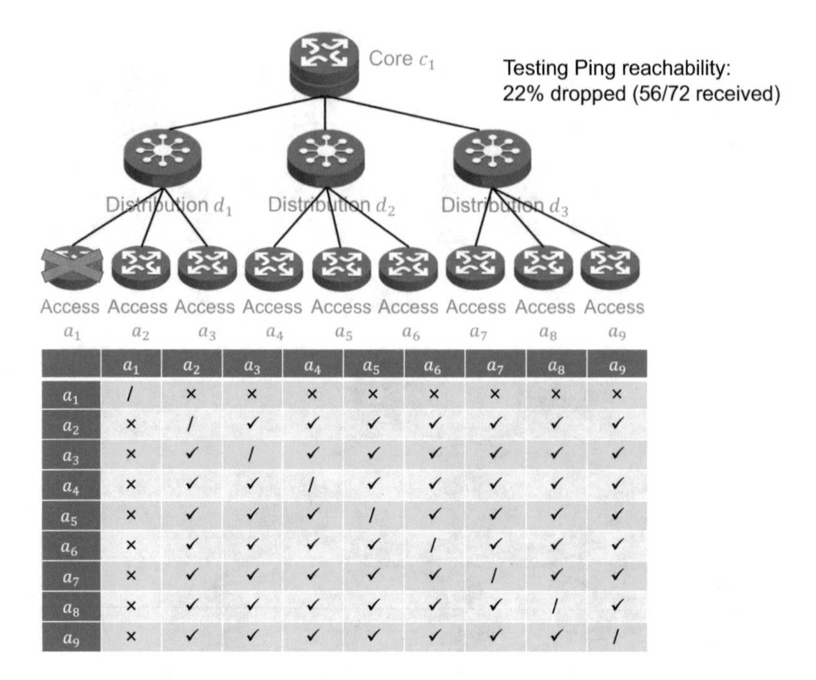

	a_1	a_2	a_3	a_4	a_5	a_6	a_7	a_8	a_9
a_1	/	×	×	×	×	×	×	×	×
a_2	×	/	✓	✓	✓	✓	✓	✓	✓
a_3	×	✓	/	✓	✓	✓	✓	✓	✓
a_4	×	✓	✓	/	✓	✓	✓	✓	✓
a_5	×	✓	✓	✓	/	✓	✓	✓	✓
a_6	×	✓	✓	✓	✓	/	✓	✓	✓
a_7	×	✓	✓	✓	✓	✓	/	✓	✓
a_8	×	✓	✓	✓	✓	✓	✓	/	✓
a_9	×	✓	✓	✓	✓	✓	✓	✓	/

Fig. 1.3 The effect of a single failure in an access node

The core layer is a key element of any telecommunication network. It should not only provide high-speed switching/routing, but also provide reliability and fault tolerance. Figures 1.3 and 1.4 show why failures occurring in the Core layer will have more severe impact than failures in other layers.

From Figs. 1.3 and 1.4, we can see that for a communication system without any fault tolerant mechanisms (such as hot/cold standby), if the single failure occurs in an access node such as a1, the other access nodes can still reach each other, although with a 22% drop rate (as shown in Fig. 1.3); however, if the single failure occurs in the Core node, access nodes cannot reach each other, leading to a 100% drop rate. Therefore, the Core node can be seen as the central forwarding subsystem inside a communication system, and the reliability of the Core node must be enhanced to maintain the normal operation of the entire system. Moreover, the functionality and performance requirements of network devices used in Core nodes far exceed beyond those used in the distribution and access nodes.

1.2 Overview of Core Router Systems

The network devices (such as routers) used in Core nodes are typically complex systems that contain both software and hardware (sometimes also operating systems), and are aimed at providing reliable, scalable and high-speed networking functions

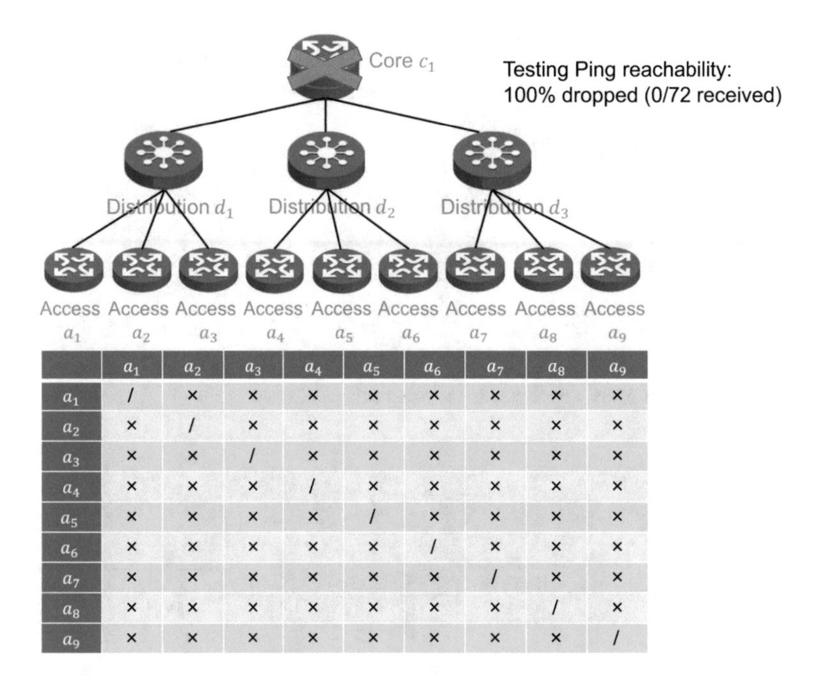

	a_1	a_2	a_3	a_4	a_5	a_6	a_7	a_8	a_9
a_1	/	×	×	×	×	×	×	×	×
a_2	×	/	×	×	×	×	×	×	×
a_3	×	×	/	×	×	×	×	×	×
a_4	×	×	×	/	×	×	×	×	×
a_5	×	×	×	×	/	×	×	×	×
a_6	×	×	×	×	×	/	×	×	×
a_7	×	×	×	×	×	×	/	×	×
a_8	×	×	×	×	×	×	×	/	×
a_9	×	×	×	×	×	×	×	×	/

Fig. 1.4 The effect of a single failure in a core node

such as switching, routing, multiplexers, cross-connects, firewalls, or load balancers [1, 2, 5]. In contrast, a network device (such as a router) used in an Access node can just be an application-specific embedded system, and it only needs to provide some basic networking functions. For instance, Fig. 1.5 shows examples of different types of routers deployed in the access node, distribution (or aggregation) node, and core node. Routers used in access nodes are commonly known as home routers and they allow home or small-business customers to connect to a wide area network (WAN) via a modem [2, 4]. Home routers are usually self-contained devices using internally stored firmware, making them simple and cheap to repair or replace [5]. Distribution routers aggregate traffic from routers in access nodes and forward them to core nodes if necessary [2]. To support a wide range of services with high quality, distribution routers usually have large memory, multiple interface connections, and onboard data processing routines, making them more powerful but also more failure-prone than home routers [4, 5]. Core routers interconnect distribution routers from multiple enterprise locations or even worldwide telecommunication providers. To provide highest bandwidth and speed in terms of transferring data packets, core routers typically are large, complex and expensive [2, 5].

This book is focused on a commercial core router system that is designed for deployment in the Internet as a super core node of the backbone network. Figure 1.6 shows an architectural overview of such a system. We can see a core router consists of a control plane and a data plane. The control plane is responsible for processing

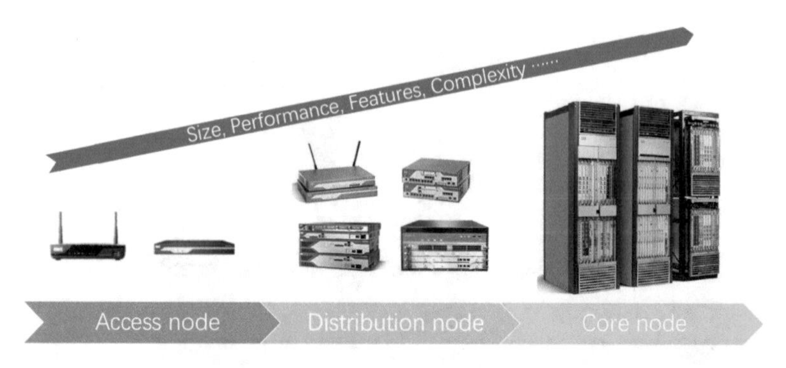

Fig. 1.5 Examples of routers deployed in different network layers

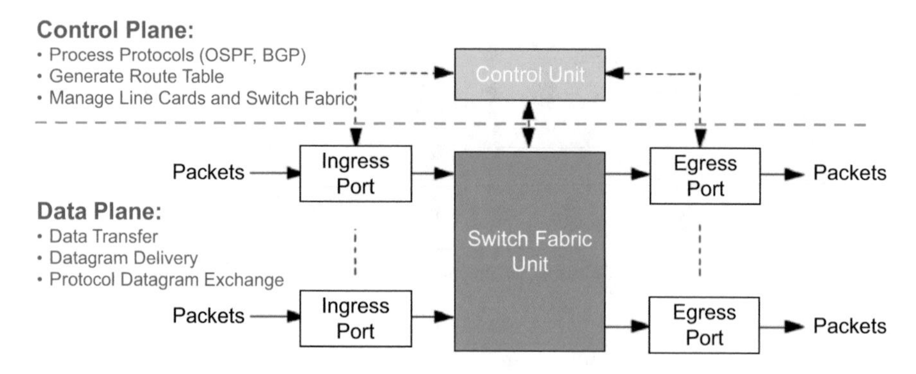

Fig. 1.6 An architectural overview of core router systems [5]

core protocols such as OSPF, BGP, etc., generating a route table, and managing all other components such as line cards, switch fabric, etc. The data plane targets data transfer between the ingress and egress ports, including datagram delivery, protocol exchange, etc.

A multi-card chassis system is commonly used in core routers. A depiction of such a multi-card chassis core router system is shown in Fig. 1.7. We can see that a multi-card chassis system consists of a number of different components such as Fan module, Power module, Line Processing Units (LPU), Switch Fabric Units (SFU), Main Processing Units (MPU), Physical Interface Cards (PIC), and so on. Each component adds specific functionalities to the system. For example, the switching fabric is a key component of a core router system. The switching fabric chip is located on the SFU; The switching fabric performs switching through cells; Upstream traffic is buffered, and downstream traffic is scheduled. The switching fabric receives data from the upstream line card, performs data switching, and then routes traffic to the corresponding downstream line card. The MPU controls and manages the system by carrying out route calculation, outband communication, device maintenance, data configuration and storage.

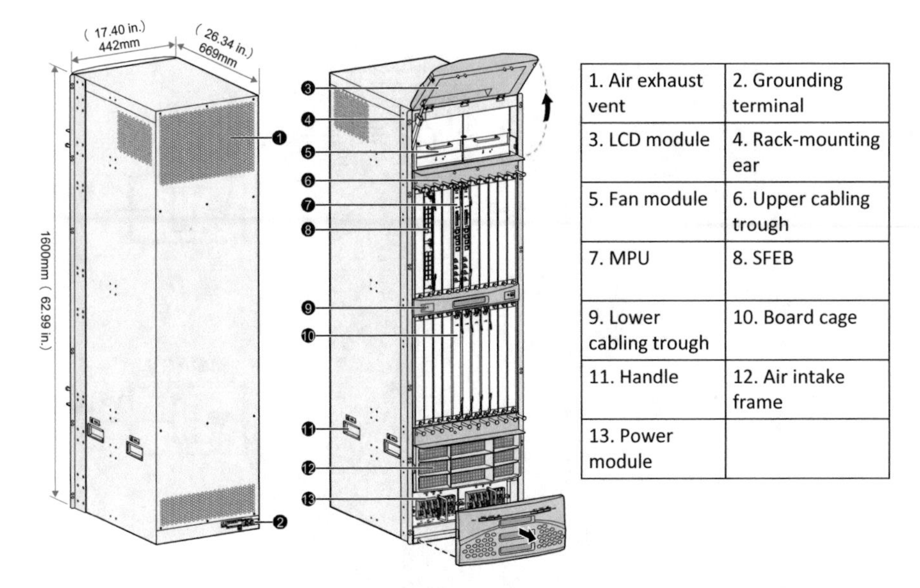

1. Air exhaust vent	2. Grounding terminal
3. LCD module	4. Rack-mounting ear
5. Fan module	6. Upper cabling trough
7. MPU	8. SFEB
9. Lower cabling trough	10. Board cage
11. Handle	12. Air intake frame
13. Power module	

Fig. 1.7 A depiction of a multi-card chassis core router system [5]

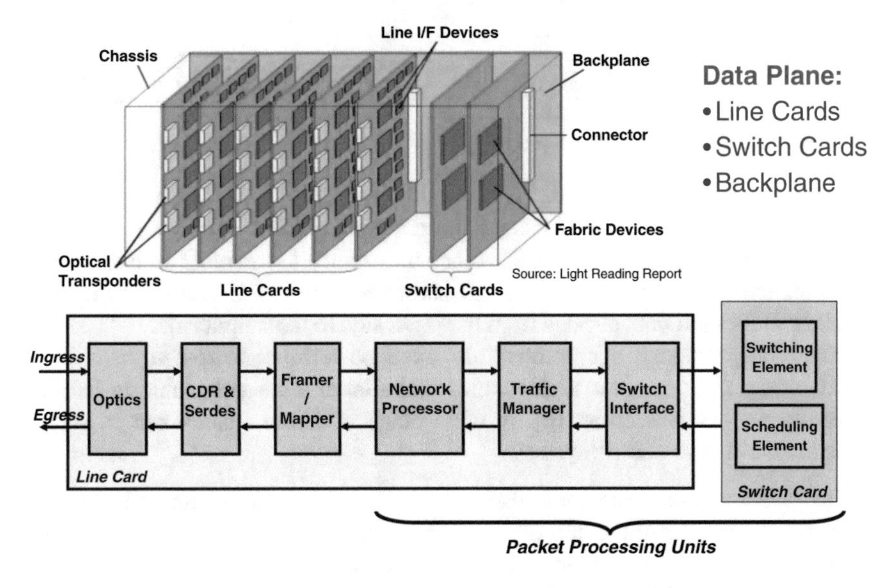

Fig. 1.8 An architecture overview of data plane in core router systems [5]

An architectural snapshot of the data plane in the multi-card chassis system is shown in Fig. 1.8. It consists of several line cards, switch cards, and a backplane to connect all the components. The line card is responsible for preprocessing and managing a large volume of data while the switch card is utilized to schedule

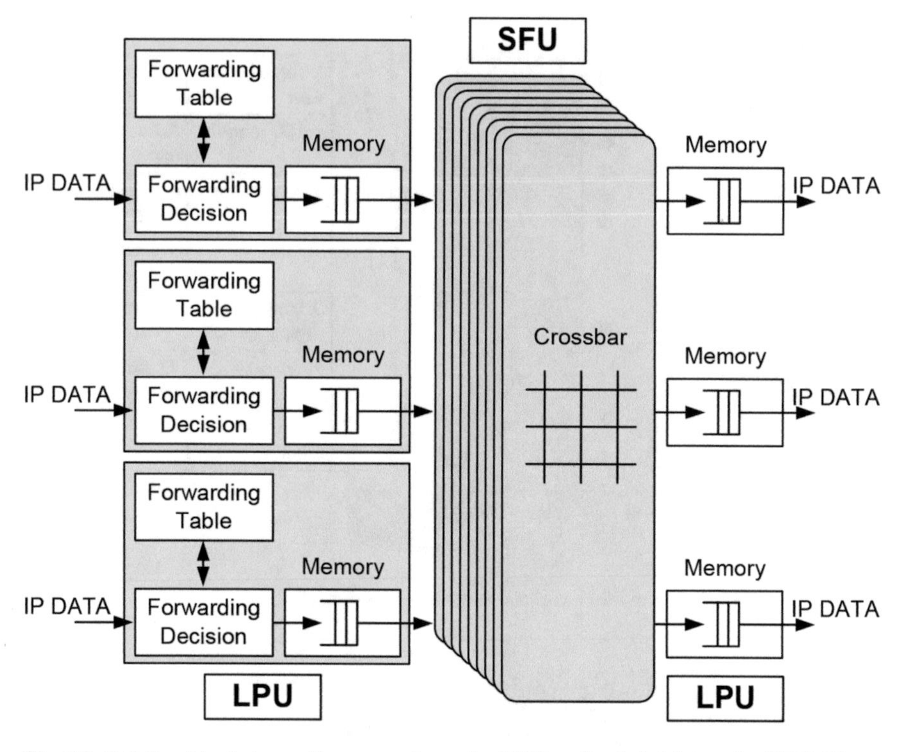

Fig. 1.9 Relationships between line processing units (LPU) and switch fabric units (SFU) [5]

and switch data from source to destination ports. In addition, different components coordinate and cooperate with each other in a concerted manner to ensure high performance and reliability. Figure 1.9 shows how the LPU and the SFU work together to forward data packets from the input side to the output side.

Therefore, although core routers are more powerful than routers used in an access node, their complex architectures make them more vulnerable to hard-to-detect/hard-to-recover errors [6]. A wide range of failures can occur in such a complex multi-card chassis system:

- Hardware failures: The cards that constitute the chassis system and the components that constitute a card can encounter hardware failures. Moreover, connectors between cards and interconnects between different components inside a card are also subject to hard faults. A multi-card chassis system can have tens of separate cards, each card can have hundreds of components, and each component consists of hundreds of advanced chips. Each chip in turn has hundreds of I/Os and millions of logic gates, and the operating frequency of chips and I/Os are now in the GHz range [5, 7]. Such high complexity and operating speed lead to an increasing number of incorrect or inconsistent hardware behaviors [8]. Moreover, in such a large-scale complex system, whenever a hardware failure occurs in the

chassis system, it is difficult for debug technicians to accurately identify the root cause of this failure and take effective corrective actions.
- Software failures: The entire chassis system and each card have their own software platforms to control and manage different network tasks. However, since the performance requirement of network devices in the core layer is approaching Tbps levels, concurrency failures caused by subtle interactions between parallel threads or applications have become more frequent because a growing number of software applications tend to distribute their tasks into parallel agents in order to improve performance [9].

All these different types of faults can cause a core router to become incapacitated, necessitating the design and implementation of fault-tolerant mechanisms for reliable computing in the core layer.

Reactive fault tolerance is commonly used to recover from failures. It aims at repair when failures actually occur on execution [10, 11]. The state-related data of a faulty system is fed to a fault-diagnosis system to identify candidate faulty components [12]. Repair actions are then executed on this list of candidates. However, such a fault-tolerance solution is only of limited applicability for today's commercial systems. A key limitation is that network devices in the core layer often require non-stop utilization (99.999% uptime) [6]; however, most reactive fault-tolerant methods need to spend a significant amount of time to identify and repair faults, which can stall system operation. In contrast, proactive fault tolerance is promising because it takes preventive action before a failure occurs, proactively repairing suspect components [10, 11]. The state of the system is monitored in a real-time manner. When unhealthy status or anomalous behaviors are identified in the system, proactive repair actions such as job migration are executed to avoid errors, thereby maintaining the non-stop utilization of the entire system [13].

1.3 Prior Work on Anomaly Detection and Health Assessment

Recent work on board-level fault diagnosis has shown that reasoning-based data-driven methods based on machine learning can greatly improve the effectiveness of fault diagnosis in reactive fault-tolerant mechanisms [7, 12, 14–17]. Although there are significant differences between reactive fault tolerance and proactive fault tolerance, proactive fault-tolerant methods can also benefit from reasoning-based data-driven techniques. As shown in Fig. 1.10, the system is monitored by recording different key performance indicators (KPIs). The logged KPI data is then fed to the health assessment component to assess the overall healthiness of current core router systems as well as detect anomalous behaviors. A failure predictor is also incorporated to identify the location and time interval for potential failures. Finally, appropriate preventive actions can be executed. The effectiveness of such data-

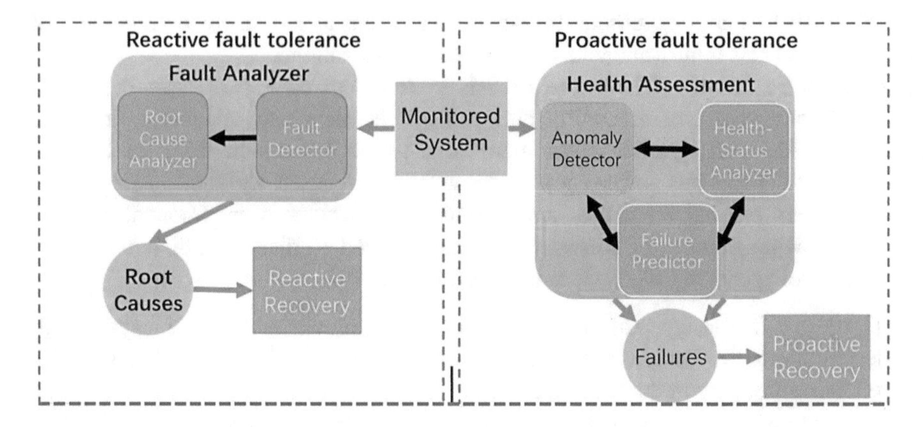

Fig. 1.10 An illustration of a data-driven fault-tolerant mechanism

driven proactive fault-tolerance solutions depends on whether unhealthy status and abnormal behaviors can be accurately pinpointed in a timely manner [18].

Anomaly detection has been widely researched in prior work to identify system's unexpected behaviors. It has been used in a variety of domains such as fraud detection for credit cards, insurance or health care, intrusion detection for cyber-security, fault detection in safety critical systems, and military surveillance for enemy activities. For example, an anomalous traffic pattern in a computer network could mean that a hacked computer is sending out sensitive data to an unauthorized destination [19]. An anomalous MRI image may indicate presence of malignant tumors [20]. Anomalies in credit card transaction data could indicate credit card or identity theft [21]. We have thus reviewed four categories of existing anomaly detection techniques [22–26].

1.3.1 Techniques Based on Statistical Models

Statistical modeling is a mathematically-formalized method that estimates the data-generating process of sample data [27]. The probability distribution P inherent in the statistical model approximates the true data distribution. Therefore, statistical models can be used for anomaly detection if the assumption that anomalous data points occur in low-probability region holds true [23]. Statistical-model-based anomaly detection usually consists of two steps. The first step is to fit a mathematical model to represent the normal behavior of sample data. The second step is to apply a statistical inference test to determine whether new data points lie in low-probability region of this statistical model [23, 28]. Both parametric and non-parametric techniques have been used for this purpose.

1.3.1.1 Parametric Techniques

Parametric techniques assume a specific probability density function $f(d, \theta)$ and estimate its parameters θ from sample data d [29]. After estimating this parametric distribution, a statistical hypothesis test is applied to test data points. The corresponding null hypothesis H_0 is that the test data point is generated using $f(d, \theta)$. The anomaly score for this test data points is thus obtained from the probability of rejecting H_0 in the hypothesis test. Two types of models are typically used in parametric techniques [23]:

1. Gaussian Model: Data points are assumed to be generated from a Gaussian distribution in Gaussian-model-based methods. The Maximum Likelihood Estimation (MLE) is widely used to estimate parameters in Gaussian models[30]. Different methods have been proposed to calculate the anomaly score based on the estimated Gaussian distribution. For example, the 3σ rule has been used to identify data points that are 3σ away from the distribution mean μ as anomalies, where σ is the standard deviation for the distribution. This is because in Gaussian distribution, the $\mu \pm 3\sigma$ region contains 99.7% of the data points [31]. The box plot rule has also been applied to detect anomalies in medical-domain data [32]. A box-plot graphically depicts groups of numerical data through their quartiles. The Inter Quartile Range (IQR) is then defined to measure the statistical dispersion of box-plot. It is calculated as $Q_3 - Q_1$, where Q_3 and Q_1 are upper and lower quartiles, respectively. Data points that lie over $1.5 \times IQR$ lower than Q_1 or over $1.5 \times IQR$ higher than Q_3 are declared as anomalies. This is because in Gaussian distribution, 99.3% of the data points are contained within $1.5 \times IQR$ boundary [33]. However, estimating a single parametric distribution is insufficient to model the normal behaviors in complex systems. Therefore, Gaussian mixture models have been proposed so that a hybrid of different Gaussian distribution is applied to identify various types of network intrusions [34].
2. Regression Model: Regression analysis is a predictive modeling process that estimates the relationships among variables. It describes how the target variable is generated from a set of predictor variables. Parameters in regression models are estimated via minimizing the difference between predicted and observed values. After fitting the regression model, statistical tests have been used to detect anomalies in test data with a certain confidence interval[35]. Different regression models have been applied to different types of data. For example, linear regression has been widely used to model linear relationship between variables [27]. Variants of linear regression model have been proposed for time-series data [27, 36]. For instance, the autoregressive (AR) model has been used to detect anomalies in IP networks [37]. In an AR model, the present/future value of a time series is linearly regressed on previous values from that same time series [38]. The autoregressive-moving-average (ARMA) model has also been used for network anomaly detection [39]. The ARMA model consists of two parts, an autoregressive (AR) part and a moving average (MA) part. The AR part does the

same job as the AR model while the MA part adds a linear combination of present and past white noise disturbance terms. The autoregressive-integrated-moving-average (ARIMA) model has been utilized to detect anomalies in multivariate time-series data [36]. The difference between the ARIMA model and the ARMA model is that the ARIMA model introduces a new part called integrated (I) part. This part is used to replace the original data value with the difference between current and previous values. This differencing process can be performed more than once to give better estimation. Robust regression has been proposed to tolerate outliers when fitting models using training dataset [40].

1.3.1.2 Non-parametric Techniques

Non-parametric techniques do not assume prior knowledge of the underlying distribution. They build statistical model and detect anomalies directly from sample data. Histogram-based methods have been widely used for intrusion detection and fraud detection [41, 42]. A histogram is an estimate of the probability distribution of sample data. It first divides the entire data range into a series of non-overlapping intervals ("bin"), and then counts how many values fall into each interval. Test data points falling outside of any intervals are identified as anomalies. Determining the optimal size of the bins in a histogram is the key to achieve both low false-alarm rate and high detection accuracy. The attribute-wise histogram method has been proposed for network intrusion detection in multivariate data [34, 43]. The histogram-based method can be considered as a simplistic kernel density estimation. Kernel functions have been widely used to approximate the probability density function in non-parametric techniques [44]. For example, the Parzen windows estimation has been applied for network intrusion detection [45, 46]. Other kernel methods have also been used to boost the performance of density estimation [47, 48]. Information-theoretic measures such as entropy and Kolomogorov Complexity have also been applied in non-parametric techniques to quantify the amount of uncertainty involved in sample data [49, 50]. Data points that induce higher irregularities in the information-theoretic measures are identified as anomalies. The local search algorithm (LSA) has been proposed to search for possible anomalies using information-theoretic measures [51].

1.3.1.3 Advantages and Disadvantages of Statistical Techniques

The advantages of using statistical models are given below [23, 28]:

1. Since statistical models are used to learn the expected behavior of the system, they do not require prior knowledge of different types of anomalies. Moreover, non-parametric techniques do not make any assumptions about the underlying statistical distribution of sample data.

2. Statistical techniques can operate in a semi-supervised setting. Anomalies in the training dataset are not required to be explicitly labeled.
3. Statistical models can not only identify outliers, but also provide additional diagnostic information such as the confidence interval for each test instance.

The disadvantages of statistical approaches include the following [23, 28]:

1. The effectiveness of statistical approaches depends on whether the expected behavior of the system can be accurately modeled by statistical distributions. However, normal behaviors in high-performance complex system can change as time proceeds and cannot be modeled by a particular distribution.
2. Tunning parameters and choosing the fittest hypothesis testing statistics are challenging [52].
3. The density estimation in statistical modeling is time consuming for multivariate data.

1.3.2 Methods Based on Clustering

Clustering is an unsupervised learning process that divides a dataset into groups such that: (1) similar objects are grouped into the same cluster; (2) dissimilar objects are located in different clusters [53, 54]. Clustering can be used for outlier detection if one of the following assumptions hold true [22, 23]:

1. Anomalous data points do not belong to any clusters or lie far away from any cluster centroid.
2. Anomalous data points are grouped into small or sparse clusters.

Most existing clustering-based anomaly detection rely on the first assumption. For example, the k-means clustering and k-medoids clustering have been used to identify anomalous network traffic [55, 56]. They first divide data points into k disjoint clusters. The distance of data point to its nearest cluster centroid is then calculated as its anomaly score. Expectation Maximization (EM) clustering has also been used for intrusion detection [56]. Instead of assigning data points to dedicated clusters, this method outputs probability of membership for each data instance. Density-based spatial clustering of applications with noise (DBSCAN) and hierarchical clustering have also been investigated because they do not require the number of clusters as prior knowledge [54, 57–59]. Moreover, in the DBSCAN algorithm, data points that are not reachable from any other points are automatically identified as outliers. The FindOut algorithm has been proposed to iteratively remove clusters and identify the final remaining data points as anomalies [60]. A disadvantage of above techniques is that they are not able to detect anomalies if these anomalies form clusters by themselves.

Methods based on the second assumption have thus been used to overcome this disadvantage. The local outlier factor (LOF) has been proposed to define the outlierness of each data point based on the density of its neighborhood [61].

Data points with low-density neighborhood (sparse cluster) tend to have high LOF values. The MINDS system has been developed based on LOF for detecting network intrusions [62]. The Cluster-Based Local Outlier Factor (CBLOF) has been proposed later to further improve the effectiveness of LOF [63]. The CBLOF takes both the density of neighborhood and the size of cluster into consideration such that data points lying in sparse or small clusters are identified as anomalies.

1.3.2.1 Advantages and Disadvantages of Clustering Methods

The advantages of using clustering are given below [23, 28]:

1. Since clustering is an unsupervised learning process, clustering-based anomaly detection does not require fully-labeled dataset, which is difficult to obtain in many cases.
2. Since clustering does not require prior knowledge of the underlying data distribution, the performance of clustering-based method is usually more stable than statistical approaches.
3. Since test data points only need to be compared with characteristics of obtained clusters, the testing time of clustering-based method is significantly reduced.

The disadvantages of clustering-based methods include [23, 28]:

1. The effectiveness of clustering-based anomaly detection depends on whether the above-mentioned two assumptions hold true. However, anomalous behaviors in high-performance complex system may not obey these assumptions.
2. Choosing fittest clustering techniques and finding optimal combination of parameters are time-consuming and error-prone.
3. Dynamic updating of clusters using new data points is difficult and computational-expensive.

1.3.3 Methods Based on Classification

Classification is a supervised learning process that identifies the categories of new observations on the basis of a labeled training dataset [64]. Classification can be used for anomaly detection if the classifier can distinguish between normal and abnormal class labels in the training dataset [23]. If instances in the training dataset only have two class labels to indicate whether they are anomalous or not, binary classification techniques can be used. Otherwise, multi-class classification techniques are needed to distinguish between each normal class and each abnormal class. A wide range of machine-learning techniques including some up-to-date deep-learning approaches have been used for this purpose.

The naive Bayes network (NB) is a probabilistic classifier based on applying Bayes' theorem [65]. The objective of NB is to build a probabilistic graphic model to represent relationship among system variables. The conditional probability model and prior class probabilities are first estimated from the training dataset, and are then used to calculate posterior. Test instances will choose the class label with largest posterior as their predicted class. Variations of NB have been used for intrusion detection on traffic bursts [66, 67].

The support vector machine (SVM) has also been widely used for anomaly detection and intrusion detection [68–71]. The objective of SVM is to find the optimal separating hyper-plane in the higher-dimensional feature space [72]. For binary-classification-based anomaly detection, test instances lying in one side of the hyper-plane are declared as normal and test instances lying in the other side are declared as anomalous. Kernels such as the radial basis function (RBF) kernel can be used in SVM to learn complex non-linear decision boundaries [73]. The one-versus-all or one-versus-one mechanism can be used in SVM to handle multi-class classification problems.

The decision tree (DT) is a tree-like predictive model and has been used for novel and intrusion detection [74, 75]. In DT model, the leaf nodes represent class labels and the branches represent decision paths [76]. Test instances start from the root node of the decision tree, and move through the internal nodes of the tree until reaching a leaf node. The class labels of test instances are then obtained from the leaf nodes. Different algorithms such as ID3, C4.5 and CART have been developed for DT-based classification [77, 78]. Variants of DT have been proposed to outperform NB approach in intrusion detection systems [79].

The artificial neural network (ANN) is a collection of interconnected units that imitates the neurons in human brain [80]. The ANN model consist of neurons and weighted connections between neurons. Neurons are arranged in layers, and weighted connections link the neurons in different layers. The back-propagation algorithm has been widely used for training ANNs [14]; it minimizes the difference between actual outputs of ANNs and the desired values using gradient descent. Test instances are fed as inputs to the ANN model and their class labels can be obtained from the output of the last layer of the ANN model. Several intrusion detection systems have been developed based on the ANN model [71, 81].

The deep neural network (DNN) can be considered as ANN with multiple hidden layers [82]. The extra layers in DNN enable composition of features from lower layers. Therefore, the DNN model can model much more complex non-linear relationships than a shallow ANN model. Variants of DNN have been proposed for network intrusion detection and anomaly detection in sequence [83–85]. For example, the recurrent neural network (RNN) is a class of DNN where connections between units form a directed cycle [86]. Recurrent neural networks can make use of the internal state to process relevant data in arbitrary sequences of inputs [83]. The autoencoder is another type of DNN that learns the generative models of sample data [87]. The autoencoder usually has the same number of input and output neurons. The objective of training autoencoder is to reconstruct its own inputs with minimum reconstruction errors. The reconstruction errors of test instances indicate whether they are anomalous or not [84].

1.3.3.1 Advantages and Disadvantages of Classification Methods

The advantages of using classification are given below [23, 28]:

1. Classification-based techniques usually have higher detection accuracy and lower false-alarm rate.
2. Classification-based techniques do not require prior knowledge of the underlying data distribution.
3. Since each test instance only needs to be fed to the precomputed classifier, the testing phase of classification-based techniques is usually fast. Moreover, many classification techniques can operate in online manner, enabling them to update their models with less effort.

The drawbacks of classification-based methods include [23, 28]:

1. Since classification is a supervised learning process, classification-based anomaly detection usually requires fully-labeled dataset for training, which is not feasible in many cases.
2. Choosing fittest classification approaches and tuning parameters are time-consuming and resource-intensive.
3. Most classification-based methods cannot identify new anomalies that are not defined in the training dataset. Moreover, they provide little diagnostic information for detected anomalies.

1.4 Research Challenges and Motivation

As shown in Fig. 1.11, a distributed agent-based platform is required to monitor a huge number of commercial core routers. Specifically, monitoring and collecting commands are sent by the data server to a number of distributed agent. Each agent then applies rule-based anomaly analysis to its core routers. As long as any anomalies are detected, the corresponding information is reported back to the data server and displayed on the web server for further root cause analysis. However, since rule-based methods rely on expert experience and do not fully utilize the collected data, a step 2.5 is added so that machine-learning techniques can help us conduct data-driven health assessment, fault diagnosis and failure prediction for core routers in a more efficient and automatic way. Finally, error recovery actions are executed via agents on suspect core routers to repair failures. Note that, domain knowledge of experts is incorporated into this platform via the configuration management component, specifying targeted failures, alert severity, analysis granularity, cost budgets, etc.

The difficulty of developing an efficient anomaly detector and health analyzer in the data-driven analysis component of the framework in Fig. 1.11 can be attributed to several reasons:

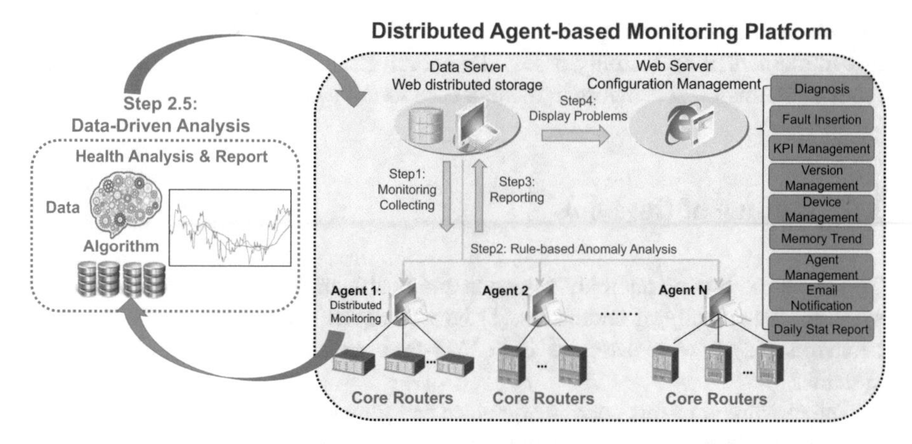

Fig. 1.11 The design of data-driven health monitoring platform for commercial core routers

1. Features extracted from communication systems are far more complex than those from a general computing system. For example, a multi-card chassis core router system uses monitors to log a large amount of features from different functional units. Each of these features can have significantly different statistical characteristics. Moreover, various types of correlation exist among different features. Therefore, it is difficult for a single type of anomaly-detection technique to be effective.
2. The monitored data in communication systems involves temporal measurements. Such time-series data can cause two problems:

    (a) The temporal dimension of the data keeps increasing as time proceeds. However, the efficiency of traditional anomaly-detection methods degrades significantly in high-dimensional data space;
    (b) New normal/abnormal patterns can appear as time proceeds and the statistical characteristics of features can change significantly even if no anomalies occur. However, traditional anomaly-detection techniques make the assumption that the statistical properties of normal features remain constant across the entire temporal domain.

3. An efficient anomaly detector is not adequate to obtain a full picture of the health status of monitored core routers. Learning different normal patterns is also important because it can reveal how healthy a core router system is and how different task scenarios can affect the system. However, two challenges exist for designing an efficient health analyzer:

    (a) Identifying a wide range of health status from long-term complex time-series data suffers from the phenomenon of "curse of dimensionality";
    (b) Labeled historical operation data are difficult to obtain in the early stages of monitoring.

Therefore, efficient anomaly detection and health status analysis need to be designed and implemented to not only assess the overall health, but also identify different types of anomalous behaviors in modern core router systems.

1.5 Outline of the Book

This book focuses on anomaly detection and health analysis in modern core router systems. Specifically, it addresses: (1) how to detect various types of anomalous behaviors; (2) how to assesses both long-term and short-term health status of systems.

The remainder of this book is organized as follows.

Chapter 2 presents an anomaly detector for core router systems using correlation-based time series analysis. The proposed technique monitors a set of features of a complex core router system. Various types of correlations among extracted features are identified. A set of features with minimum redundancy and maximum relevance are then grouped into different categories based on their statistical characteristics. A hybrid approach is developed to analyze different feature categories with different anomaly detection methods, leading to higher success ratios for detecting different types of anomalies.

Chapter 3 presents the design of a changepoint-based anomaly detector that first detects changepoints from collected time-series data, and then utilizes these changepoints to detect anomalies. Different changepoint detection approaches are implemented to detect different types of changepoints. A clustering method is then developed to identify a wide range of normal/abnormal patterns from changepoint windows. Experimental results show that our changepoint-based anomaly detector can detect "outliers" even when the statistical properties of the monitored data change significantly as time proceeds.

A symbol-based health status analyzer is presented in Chap. 4 to obtain a full picture of the health status of monitored core routers. The proposed method first encodes, as a symbol sequence, the long-term complex time series collected from a number of core routers, and then utilizes the symbol sequence to do health analysis. Multiple symbolization techniques are implemented to encode complex time series in a hierarchical way. Several symbol-based clustering and classification methods are developed to identify the health status of core routers. Experimental results show that the proposed symbol-based method can maintain its effectiveness as the length of time series increases.

Chapter 5 is focused on a self-learning health analyzer for partially-labeled data extracted from core router systems. The proposed method first computes a representative feature matrix to capture different characteristics of time-series data. Hierarchical clustering is then utilized to infer labels for the unlabeled dataset. Finally, a classifier is built and iteratively updated using both labeled and unlabeled dataset. Partially-labeled field data collected from a set of commercial core routers are used to experimentally validate the proposed method.

Finally, Chap. 6 summarizes the contributions of the book and identifies directions for future work.

References

1. P. Veitch et al., Integrating core BGP/MPLS networks. Internet Protoc. J. **13**(4), 18–31 (2010)
2. C.N. Academy, *Connecting Networks Companion Guide* (Pearson Education, London, 2014)
3. V. Antonenko, R. Smelyanskiy, Global network modelling based on mininet approach, in *Proceedings of the Second ACM SIGCOMM Workshop on Hot Topics in Software Defined Networking* (2013), pp. 145–146
4. L. Wayne, *LAN Switching and Wireless, CCNA Exploration Companion Guide* (Pearson Education India, Delhi, 2008)
5. R. Giladi, *Network Processors: Architecture, Programming, and Implementation* (Morgan Kaufmann, San Francisco, 2008)
6. M. Médard, S.S. Lumetta, Network reliability and fault tolerance, in *Encyclopedia of Telecommunications* (Wiley, New York, 2003)
7. S. Jin, F. Ye, Z. Zhang, K. Chakrabarty, X. Gu, Efficient board-level functional fault diagnosis with missing syndromes. IEEE Trans. Comput. Aided Des. Integr. Circuits Syst. **35**(6), 985–998 (2016)
8. R. Isermann, *Fault-Diagnosis Systems: An Introduction from Fault Detection to Fault Tolerance* (Springer Science & Business Media, Berlin, 2006)
9. B. Schroeder, G.A. Gibson, A large-scale study of failures in high-performance computing systems. IEEE Trans. Dependable Secure Comput. **7**, 337–350 (2010)
10. P.K. Patra, H. Singh, G. Singh, Fault tolerance techniques and comparative implementation in cloud computing. Int. J. Comput. Appl. **64**, 37–41 (2013)
11. P.A. Lee, T. Anderson, *Fault Tolerance: Principles and Practice*, vol. 3 (Springer Science & Business Media, Berlin, 2012)
12. F. Ye, Z. Zhang, K. Chakrabarty, X. Gu, Information-theoretic syndrome and root-cause analysis for guiding board-level fault diagnosis, in *Proceedings of IEEE European Test Symposium (ETS)* (2013), pp. 1–6
13. C. Wang, F. Mueller, C. Engelmann, S.L. Scott, Proactive process-level live migration in HPC environments, in *Proceedings of the 2008 ACM/IEEE Conference on Supercomputing* (2008), p. 43
14. F. Ye, Z. Zhang, K. Chakrabarty, X. Gu, Board-level functional fault diagnosis using artificial neural networks, support-vector machines, and weighted-majority voting. IEEE Trans. Comput. Aided Des. Integr. Circuits Syst. **32**, 723–736 (2013)
15. F. Ye, Z. Zhang, K. Chakrabarty, X. Gu, Knowledge discovery and knowledge transfer in board-level functional fault diagnosis, in *Proceedings of IEEE International Test Conference (ITC)* (2014), pp. 1–10
16. C.-K. Hsu et al., Test data analytics–exploring spatial and test-item correlations in production test data, in *Proceedings of IEEE International Test Conference (ITC)* (2013), pp. 1–10
17. S. Tanwir et al., Information-theoretic and statistical methods of failure log selection for improved diagnosis, in *Proceedings of ITC*, 2015
18. A. Gainaru, F. Cappello, M. Snir, W. Kramer, Fault prediction under the microscope: a closer look into hpc systems, in *Proceedings of the International Conference on High Performance Computing, Networking, Storage and Analysis* (2012), p. 77
19. V. Kumar, Parallel and distributed computing for cybersecurity. IEEE Distrib. Syst. Online **6**, 1–9 (2005)
20. C. Spence, L. Parra, P. Sajda, Detection, synthesis and compression in mammographic image analysis with a hierarchical image probability model, in *Proceedings of the IEEE Workshop on Mathematical Methods in Biomedical Image*, 2001

21. E. Aleskerov, B. Freisleben, B. Rao, Cardwatch: a neural network based database mining system for credit card fraud detection, in *Proceedings of the IEEE/IAFE Computational Intelligence for Financial Engineering* (1997), pp. 220–226
22. A. Patcha, J.-M. Park, An overview of anomaly detection techniques: existing solutions and latest technological trends. Comput. Netw. **51**, 3448–3470 (2007)
23. V. Chandola, A. Banerjee, V. Kumar, Anomaly detection: a survey. ACM Comput. Surv. **15:1–15:58**, 15 (2009)
24. P. Gogoi et al., A survey of outlier detection methods in network anomaly identification. Comput. J. **54**, 570–588 (2011)
25. S. Agrawal, J. Agrawal, Survey on anomaly detection using data mining Techniques. Proc. Comput. Sci. **60**, 708–713 (2015)
26. B. Al-Musawi et al., BGP anomaly detection techniques: a survey. IEEE Commun. Surv. **19**, 377–396 (2017)
27. S. Konishi, G. Kitagawa, *Information Criteria and Statistical Modeling* (Springer Science & Business Media, Berlin, 2008)
28. M.H. Bhuyan et al., Network anomaly detection: methods, systems and tools. IEEE Commun. Surv. Tutorials **16**, 303–336 (2014)
29. E. Eskin, Anomaly detection over noisy data using learned probability distributions, in *Proceedings of the Seventeenth International Conference on Machine Learning* (2000), pp. 255–262
30. J. Aldrich, R.a. fisher and the making of maximum likelihood 1912–1922. Stat. Sci. **12**, 162–176 (1997)
31. E. Grafarend, *Linear and Nonlinear Models: Fixed Effects, Random Effects, and Mixed Models* (Walter de Gruyter, Berlin, 2006)
32. H.E. Solberg, A. Lahti, Detection of outliers in reference distributions: performance of Horn's algorithm. Clin. Chem. **51**, 2326–2332 (2005)
33. M. Hubert, E. Vandervieren, An adjusted boxplot for skewed distributions. Comput. Stat. Data Anal. **52**, 5186–5201 (2008)
34. K. Yamanishi et al., On-line unsupervised outlier detection using finite mixtures with discounting learning algorithms. Data Min. Knowl. Discov. **8**, 275–300 (2004)
35. P.H. Torr, D.W. Murray, Outlier detection and motion segmentation, in *Sensor Fusion VI*, vol. 2059 (1993), pp. 432–444
36. P. Galeano, D. Peña, R.S. Tsay, Outlier detection in multivariate time series by projection pursuit. J. Am. Stat. Assoc. **101**, 654–669 (2006)
37. M. Thottan, C. Ji, Anomaly detection in IP networks. IEEE Trans. Signal Process. **51**(8), 2191–2204 (2003)
38. V. Vapnik, An overview of statistical learning theory. IEEE Trans. Neural Netw. **10**(5), 988–999 (1999)
39. M. Celenk, T. Conley, J. Willis, J. Graham, Predictive network anomaly detection and visualization. IEEE Trans. Inf. Forensics Secur. **5**, 288–299 (2010)
40. P.J. Rousseeuw, A.M. Leroy, *Robust Regression and Outlier Detection*, vol. 589 (Wiley, New York, 2005)
41. E. Eskin, A. Arnold, M. Prerau, L. Portnoy, S. Stolfo, A geometric framework for unsupervised anomaly detection, in *Applications of Data Mining in Computer Security* (Springer, Berlin, 2002), pp. 77–101
42. T. Fawcett, F. Provost, Activity monitoring: noticing interesting changes in behavior, in *Proceedings of the Fifth ACM SIGKDD International Conference on Knowledge Discovery and Data Mining* (1999), pp. 53–62
43. L.L. Ho, C.J. Macey, R. Hiller, A distributed and reliable platform for adaptive anomaly detection in IP networks, in *International Workshop on Distributed Systems: Operations and Management* (1999), pp. 33–46
44. B.W. Silverman, *Density Estimation for Statistics and Data Analysis* (Routledge, London, 2018)

45. E. Parzen, On estimation of a probability density function and mode. Ann. Math. Stat. **33**, 1065–1076 (1962)
46. D.-Y. Yeung, C. Chow, Parzen-window network intrusion detectors, in *16th International Conference on Pattern Recognition, 2002. Proceedings*, vol. 4 (2002), pp. 385–388
47. C.S. Teh, C.P. Lim, Monitoring the formation of kernel-based topographic maps in a hybrid SOM-kMER model. IEEE Trans. Neural Netw. **17**, 1336–1341 (2006)
48. G. Bloch et al., Reduced-size kernel models for nonlinear hybrid system Identification. IEEE Trans. Neural Netw. **22**, 2398–2405 (2011)
49. W. Lee, D. Xiang, Information-theoretic measures for anomaly detection, in *2001 IEEE Symposium on Security and Privacy, 2001. S&P 2001. Proceedings* (2001), pp. 130–143
50. L. Ming, P. Vitányi, *An Introduction to Kolmogorov Complexity and Its Applications* (Springer, Heidelberg, 1997)
51. Z. He, S. Deng, X. Xu, J.Z. Huang, A fast greedy algorithm for outlier mining, in *Advances in Knowledge Discovery and Data Mining* (2006), pp. 567–576
52. H. Motulsky, *Intuitive Biostatistics: A Nonmathematical Guide to Statistical Thinking* (Oxford University Press, Oxford, 2013)
53. P.-N. Tan, M. Steinbach, V. Kumar et al., *Introduction to Data Mining* (Pearson Education India, Delhi, 2006)
54. O. Maimon, L. Rokach, *Data Mining and Knowledge Discovery Handbook* (Springer, New York, 2005)
55. G. Münz, S. Li, G. Carle, Traffic anomaly detection using k-means clustering, in *GI/ITG Workshop MMBnet*, 2007
56. I. Syarif, A. Prugel-Bennett, G. Wills, Data mining approaches for network intrusion detection: from dimensionality reduction to misuse and anomaly detection. J. Inf. Technol. Rev. **3**, 70–83 (2012)
57. M. Ester, H.-P. Kriegel, J. Sander, X. Xu, et al., A density-based algorithm for discovering clusters in large spatial databases with noise, in *Kdd*, vol. 96, no. 34 (1996), pp. 226–231. https://scholar.googleusercontent.com/scholar.bib?q=info:-KybkyxcGYIJ:scholar.google.com/&output=citation&scisdr=CgXKeW0REOXD7i92Y3Q:AAGBfm0AAAAAXe1ze3R8hFXHZTU1cF2vgN_fTvfsj7n8&scisig=AAGBfm0AAAAAXe1ze25z3yqnbvDqkrKyrFQlrHSHZvKk&scisf=4&ct=citation&cd=-1&hl=en
58. S. Guha, R. Rastogi, K. Shim, Rock: a robust clustering algorithm for categorical attributes, in *15th International Conference on Data Engineering, 1999. Proceedings* (1999), pp. 512–521
59. L. Ertöz, M. Steinbach, V. Kumar, Finding topics in collections of documents: a shared nearest neighbor approach, in *Clustering and Information Retrieval* (Springer, 2004), pp. 83–103. https://scholar.googleusercontent.com/scholar.bib?q=info:6_FGDFfMgdsJ:scholar.google.com/&output=citation&scisdr=CgXKeW0REOXD7i92rAo:AAGBfm0AAAAAXe1ztApZ5-i7PHwRAvc7Jl9hQS2iKb77&scisig=AAGBfm0AAAAAXe1ztPaMM5l8A2fOfe5_nNBur3GiTTPt&scisf=4&ct=citation&cd=-1&hl=en
60. D. Yu, G. Sheikholeslami, A. Zhang, Findout: finding outliers in very large datasets, in *Knowledge and Information Systems*, vol. 4 (Springer, London, 2002), pp. 387–412
61. M.M. Breunig et al., Lof: identifying density-based local outliers, in *ACM Sigmod Record*, vol. 29 (ACM, New York, 2000), pp. 93–104
62. L. Ertoz et al., Minds-Minnesota intrusion detection system, in *Next Generation Data Mining* (MIT Press, Boston, 2004), pp. 199–218
63. Z. He, X. Xu, S. Deng, Discovering cluster-based local outliers. Pattern Recogn. Lett. **24**, 1641–1650 (2003)
64. E. Alpaydin, *Introduction to Machine Learning* (MIT Press, Cambridge, 2014)
65. S.J. Russell, P. Norvig, *Artificial Intelligence: A Modern Approach* (Pearson Education Limited, Malaysia/Prentice Hall, Englewood Cliffs, 2016). https://scholar.googleusercontent.com/scholar.bib?q=info:I5nM5aK3CioJ:scholar.google.com/&output=citation&scisdr=CgXKeW0REOXD7i9xxp0:AAGBfm0AAAAAXe103p0rY6JwTpsa52HsHX_I0dv1R3fl&scisig=AAGBfm0AAAAAXe103pLBwolLgLFeP9K3udaS0GJ3tkhr&scisf=4&ct=citation&cd=-1&hl=en

66. C. Kruegel, D. Mutz, W. Robertson, F. Valeur, Bayesian event classification for intrusion detection, in *Computer Security Applications Conference, 2003. Proceedings. 19th Annual* (2003), pp. 14–23
67. D. Janakiram, V. Adi Mallikarjuna Reddy, A. Kumar, Outlier detection in wireless sensor networks using Bayesian belief networks, in *International Conference on Communication System Software and Middleware* (2006), pp. 1–6
68. G. Rätsch, S. Mika, B. Scholkopf, K.-R. Müller, Constructing boosting algorithms from SVMs: an application to one-class classification. IEEE Trans. Pattern Anal. Mach. Intell. **24**, 1184–1199 (2002)
69. K.A. Heller et al., One class support vector machines for detecting anomalous windows registry accesses, in *Proceedings of the Workshop on Data Mining for Computer Security*, vol. 9, 2003
70. A. Lazarevic et al., A comparative study of anomaly detection schemes in network intrusion detection, in *Proceedings of the 2003 SIAM International Conference on Data Mining* (2003), pp. 25–36
71. S. Mukkamala, G. Janoski, A. Sung, Intrusion detection using neural networks and support vector machines, in *Proceedings of the International Joint Conference on Neural Networks*, vol. 2 (2002), pp. 1702–1707
72. C. Cortes, V. Vapnik, Support-vector networks. Mach. Learn. **20**, 273–297 (1995)
73. D.M.J. Tax, One-class classification: concept-learning in the absence of counter-examples. PhD thesis, Delft University of Technology, 2001
74. W. Lee et al., A data mining and CIDF based approach for detecting novel and distributed intrusions, in *International Workshop on Recent Advances in Intrusion Detection* (2000), pp. 49–65
75. W. Lee, S.J. Stolfo, K.W. Mok, Adaptive intrusion detection: a data mining approach. Artif. Intell. Rev. **14**, 533–567 (2000)
76. J. Quinlan, Induction of decision trees. Mach. Learn. **1**, 81–106 (1986)
77. J.R. Quinlan et al., Bagging, boosting, and c4. 5, in *AAAI/IAAI, Vol. 1* (1996), pp. 725–730
78. R.L. Lawrence, A. Wright, Rule-based classification systems using classification and regression tree (cart) analysis. Photogramm. Eng. Remote Sens. **67**, 1137–1142 (2001)
79. N.B. Amor, S. Benferhat, Z. Elouedi, Naive Bayes vs decision trees in intrusion detection systems, in *Proceedings of the 2004 ACM Symposium on Applied Computing* (2004), pp. 420–424
80. L.V. Fausett et al., *Fundamentals of Neural Networks: Architectures, Algorithms, and Applications*, vol. 3 (Prentice-Hall, Englewood Cliffs, 1994)
81. M. Amini, R. Jalili, H.R. Shahriari, Rt-unnid: a practical solution to real-time network-based intrusion detection using unsupervised neural networks. Comput. Secur. **25**, 459–468 (2006)
82. J. Schmidhuber, Deep learning in neural networks: an overview. Neural Netw. **61**, 85–117 (2015)
83. M.S. alDosari, Unsupervised anomaly detection in sequences using long short term memory recurrent neural networks, Master's thesis, 2016
84. N. Shone et al., A deep learning approach to network intrusion detection. IEEE Trans. Emerg. Top. Comput. Intell. **2**, 41–50 (2018)
85. W. Wang et al., Hast-ids: learning hierarchical spatial-temporal features using deep neural networks to improve intrusion detection. IEEE Access **6**, 1792–1806 (2018)
86. A. Graves et al., Speech recognition with deep recurrent neural networks, in *International Conference on Acoustics, Speech and Signal Processing* (2013), pp. 6645–6649
87. G.E. Hinton, R.R. Salakhutdinov, Reducing the dimensionality of data with neural networks. Science **313**, 504–507 (2006)

Chapter 2
Accurate Anomaly Detection Using Correlation-Based Time-Series Analysis

In this chapter, we present an accurate anomaly detector for core router systems using correlation-based time-series analysis. The proposed method monitors the time-series data of a complex core router system. Anomaly detection techniques are compared in terms of their effectiveness for detecting different types of anomalies. A feature-categorizing-based hybrid method is proposed to overcome the difficulty of detecting anomalies in features with different statistical characteristics. Furthermore, a correlation analyzer is implemented to remove irrelevant and redundant features. Three types of synthetic anomalies, generated using a small amount of real data for a commercial telecom system, are used to validate the proposed anomaly detector.

The remainder of the chapter is organized as follows. Section 2.1 presents the motivation of data-driven anomaly detector for core routers. Section 2.2 describes different time-series-based anomaly detection methods in more detail. Section 2.3 presents a correlation analyzer that can select the most important features and cluster correlated features. Experimental results for a commercial core router system are presented in Sect. 2.4. Finally, Sect. 2.5 concludes the chapter.

2.1 Motivation

Anomaly detection, which is also sometimes referred to as outlier detection, has been widely used in domains—such as intrusion detection and fraud detection [1, 2]. For example, density-based techniques such as k-nearest neighbor (KNN) have been used in detecting outliers in high-dimension datasets [3]. Machine-learning methods such as artificial neural networks (ANN) have also been applied to detect fraud in large multivariate databases [4]. A multivariate state estimation technique (MSET), has been used to reduce or eliminate No-Trouble-Found [5]. This technique is sensitive to subtle changes in the signal trend, making it effective in detecting

S. Jin et al., *Anomaly-Detection and Health-Analysis Techniques for Core Router Systems*, https://doi.org/10.1007/978-3-030-33664-6_2

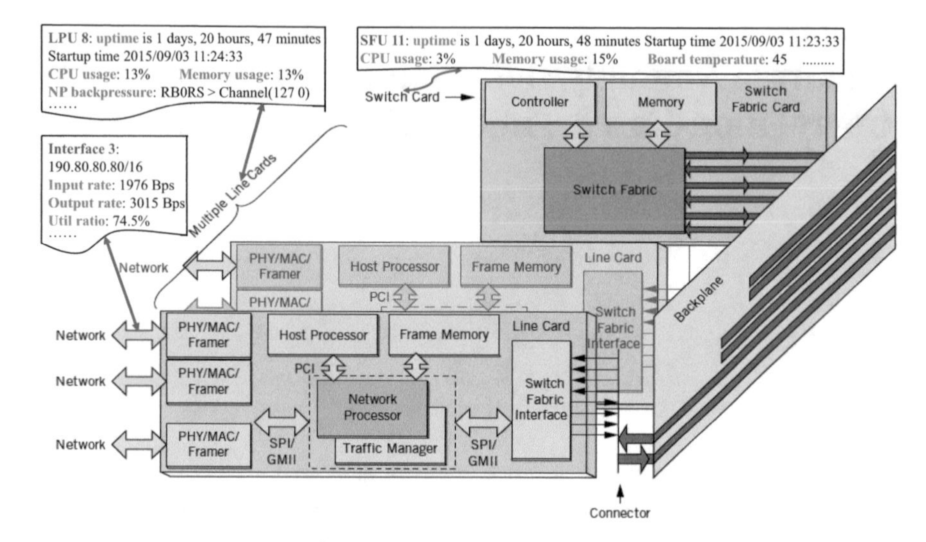

Fig. 2.1 A multi-card chassis core router system and a snapshot of extracted (monitored) features (based on [7] and data collected by the authors)

indirect anomalies. However, the non-linear operator used in MSET is essential to its performance and it is difficult to find an optimal non-linear operator for features with significantly different statistical characteristics [6].

The effectiveness of anomaly-detection techniques depends on several factors. The first among these is the dimension of feature space. The computational complexity of some methods such as ANN increases exponentially with the number of features, making them unsuitable for high-dimension datasets [4]. The second factor is the statistical characteristics of features. For example, auto regression (AR) is more effective when features are linearly correlated while cross correlation performs better when features repeat periodically [2]. The third factor is whether the data set is labeled. If a data set has been labeled as being "normal" or "abnormal", supervised classifiers can be trained to detect anomalies. Otherwise, unsupervised statistical methods must be used to find "outliers" that are the least fit to the remainder of the data set [1, 2].

The difficulty of developing an efficient anomaly detector for a complex communication system can be attributed to two reasons. The first reason is that features extracted from communication systems are far more complex than those from a general computing system. For example, as shown in Fig. 2.1, a multi-card chassis core router system uses monitors to log a large amount of features from different functional units. These features include performance metrics (e.g., events, bandwidth, throughput, latency, jitter, error rate), resource usage (e.g., CPU, memory, pool, thread, queue length), low-level hardware information (e.g., voltage, temperature, interrupts), configuration status of different network devices, and so on. Each of these features can have significantly different statistical characteristics, making it difficult for a single type of anomaly-detection technique to be effective. The second reason is that the monitored data in communication systems involves temporal measurements. Most existing anomaly-detection methods are not designed to address time-series data [1], hence they may not be able to detect time-series-specific anomalies such as the trend anomaly.

We therefore address the important practical problem of designing an anomaly detector that can be effectively applied to a commercial core router system. We use multiple anomaly-detection techniques to detect different types of anomalies. We also describe a new feature-categorization-based hybrid method to improve the performance of our anomaly detector. A correlation analyzer is also developed to select the most important features and cluster correlated features.

2.2 Time-Series-Based Anomaly Detection

In complex communication systems such as a core router, data is collected in the form of time-series. A *time series* is a series of data points indexed in time order. Formally, a multivariate time series ts consists of a set of data points in sequential order: $ts = \{tsd_1, tsd_2, \ldots, tsd_n\}$, where n is the number of data points. Each data point tsd_i consists of a set of features $tsd_i = \{f1_i, f2_i, \ldots, fv_i\}$, where v is the number of features. As shown in Fig. 2.1, these features are key performance indicators (KPIs) extracted from different functional units of core router systems, including performance metrics, resource usage, low-level hardware information, configuration status and so on. All the collected time-series data are typically divided into two datasets: the training dataset Tr and the test dataset T. A key challenge here is to detect anomalies in time-series dataset to determine whether the system is entering a degraded state or likely to fail. Therefore, we have studied a range of techniques that may be used to detect anomalies in time-series data.

2.2.1 *Distance-Based Anomaly Detection*

The illustration of "distance" between a pair of time-series instances is shown in Fig. 2.2. Assume that $y1$ and $y2$ are a pair of time series data; therefore, $y1_i$ and $y2_i$ represent the observation value of time series $y1$ and $y2$ at time point x_i. Therefore, the "distance" between $y1_i$ and $y2_i$ can be defined as $|y1_i - y2_i|$, as indicated by red double arrows in Fig. 2.2. The overall "distance" between these $y1$ and $y2$ can be calculated as $\sum_i |y1_i - y2_i|$. Such a distance measure between a pair of time-series instances can represent the similarity between these two time-series. The smaller the overall "distance" is, the closer this pair of time-series instances would be.

The k-nearest neighbors algorithm (kNN) is a widely-used method for classification and regression problem. As demonstrated in prior work [3], since a kNN is able to estimate similarities between a pair of data sets, we also utilize it for anomaly detection. Figure 2.3 illustrates kNN-based anomaly detection. Assume that the training dataset is Tr and the test dataset is T. Then for each time-series

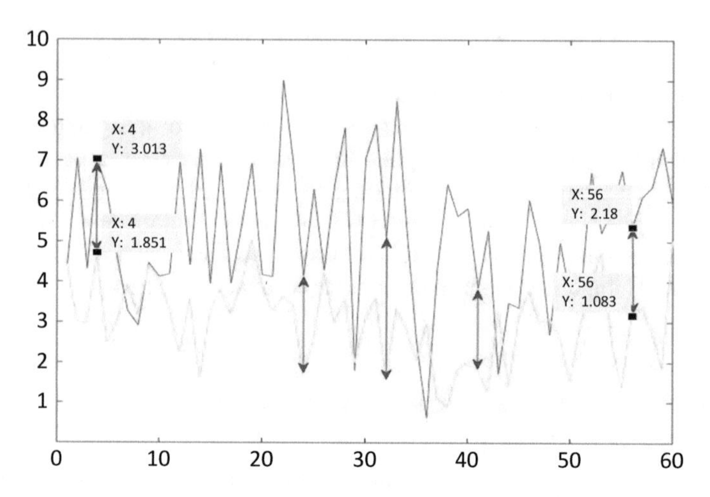

Fig. 2.2 Illustration of the "distance" between a pair of time-series instances

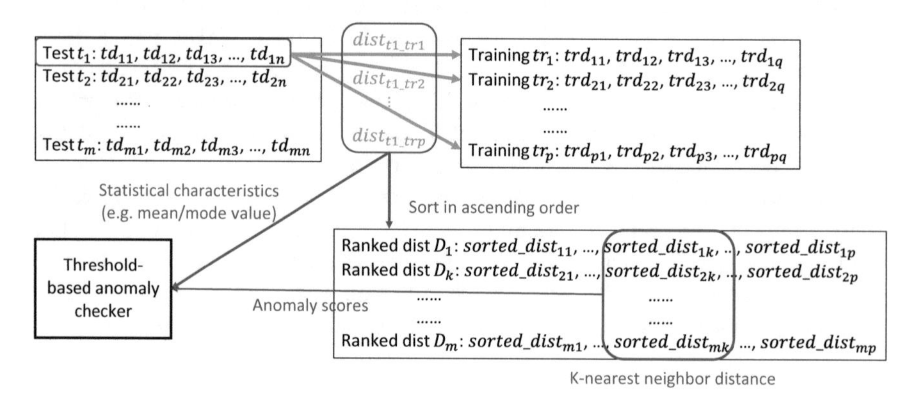

Fig. 2.3 The computation flow in KNN-based anomaly detection

instance td_i in T, we calculate its distance to all instances trd_j in Tr. The list of distances obtained in this manner are sorted in ascending order, and the kth element in the sorted distance list will be considered as the anomaly score of instance td_i. An anomaly checker is designed to ascertain whether the anomaly score of a single time-series instance significantly exceeds a predefined threshold (such as the average anomaly score). If so, this instance will be identified as being abnormal.

2.2.2 Window-Based Anomaly Detection

Window-based methods are needed to decrease the number of false alarms caused by misalignment of normal signals [1]. Figure 2.4 shows a pair of time-shifted time

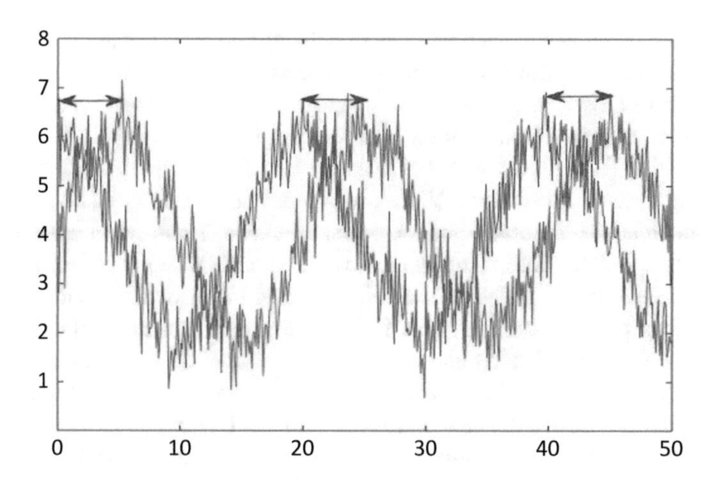

Fig. 2.4 A pair of time-shifted time series instances

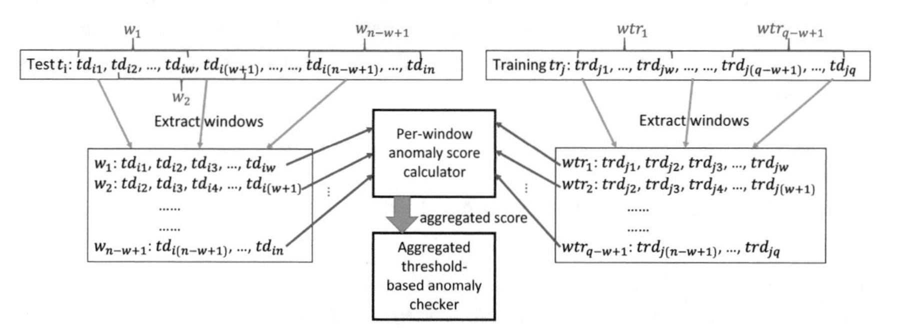

Fig. 2.5 The computation flow in window-based anomaly detection

series instances. if the "blue" wave is defined as normal data, the "red" wave should also be identified as being normal because it is just a time-shifted version of the "blue" wave. However, if KNN-based anomaly detection method is applied here, the accumulated "distance" between the "blue" and "red" wave will be high because most of the observed data points of the "red" wave are far from those of the "blue" wave. Therefore it is highly likely that the "red" wave will be incorrectly identified as being anomalous.

Consider Fig. 2.5 as an example. Assume that the length of each testing time-series instance is n, the length of each training time-series instance is q, and the length of each window is w. Then $n-w+1$ overlapping windows are extracted from each testing time-series instance and $q-w+1$ overlapping windows are extracted from each training time-series instance [2]. Then all the extracted windows are fed to a per-window anomaly score calculator.

Two types of per-window anomaly score calculation methods are investigated here. The first is based on unsupervised distance-based methods such as KNN; the anomaly score assigned to each window of a test time-series instance td_i equals the distance between the window and its kth nearest neighbor in the set of windows extracted from the training time-series instance trd_i. The second type is based on supervised prediction-based techniques. For example, as described in [2], a support-vector machine (SVM) can be utilized here. First, all the windows extracted from the training time-series dataset Tr are used to develop a one-class SVM classifier to distinguish between normal and abnormal windows. Then, when a test case t_i is obtained, it is divided into a set of windows and each window is fed to the built one-class classifier. The classifier outputs the class label for this window with the corresponding probability. The abnormality score for this window thus equals the window's probability of being classified as being abnormal. Next, the complete set of per-window scores is fed to a score aggregator to obtain the final anomaly score for each test time-series instance. Different statistical aggregation methods can be used for this purpose. For example, the aggregated anomaly score for a test time-series instance can be calculated as the mean/median/mode score value of all its windows. Finally, the aggregated anomaly scores can be fed to a threshold-based anomaly checker. Only when the anomaly score of a single time-series instance significantly exceeds a predefined threshold, this instance will be identified as being abnormal.

2.2.3 Prediction-Based Anomaly Detection

The computation flow of prediction-based anomaly detection is illustrated in Fig. 2.6. First, a predictive model is learned from historical logs. Different machine-learning techniques can be applied in this training phase. Next, test time-series instances are divided into two parts at time point t_k. The data points td_i between $[t_1,t_k]$ are fed to the predictive model, and the predictive model generates predicted data tp_i lying in the time interval $[t_{k+1},t_n]$. These predicted values are then compared with the actual measured data points td_i between $[t_{k+1},t_n]$. The accumulated difference between these predicted and the actual observations is defined as the anomaly score for each test time-series instance. Finally, an anomaly checker is designed to evaluate whether the anomaly score indicate existence of anomalies. The Autoregressive (AR) statistical forecasting model, three machine-learning techniques—Support Vector Regression (SVR), Artificial Neural Network (ANN), and Decision Tree (DT)—and a deep-learning approach, i.e., Recurrent Neural Network (RNN), are used to forecast time series and identify potential anomalies. We next describe the machine-learning techniques that we utilize for anomaly detection.

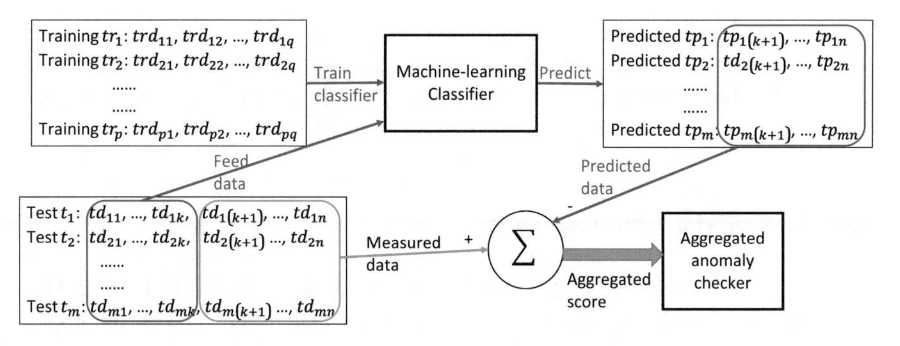

Fig. 2.6 The computation flow in prediction-based anomaly detection

2.2.3.1 Autoregressive Model

An autoregressive (AR) model is a widely used statistical forecasting model [8]. It specifies that the output variable depends linearly on its own previous values and on a stochastic term. Therefore, in an AR model, the present/future value of a time series is regressed on previous values from that same time series. The order of an AR is the number of immediately preceding values in the series that are used to predict the value at the present time. For example, a first-order AR model can be simply written as $y_t = \beta_0 + \beta_1 y_{t-1} + \epsilon_t$, where y_t is value at time t, y_{t-1} is preceding value at time $t - 1$, β_i are parameters and ϵ_t is noise term.

2.2.3.2 Support Vector Regression

A Support-Vector Machine (SVM) is a supervised machine-learning technique [9] that has been applied to various fields as a powerful classification tool. The goal of SVMs is to define an optimal separating hyperplane (OSH) to separate two classes. The vectors from the same class fall on the same side of the OSH, and the distance from the closest vectors to the OSH is the maximum among all the separating hyperplanes. Support vector regression (SVR) is a regression version of SVM, and is used to predict values that are close to true values.

2.2.3.3 Decision Tree

A decision tree (DT) is a tree-like predictive model that is widely used in statistics, data mining, and machine learning [10]. A DT consists of two types of nodes, leaf (terminal) nodes and decision (internal) nodes. Leaf nodes refer to the nodes that do not branch and contain prediction information. Decision nodes refer to the nodes that can branch to multiple child nodes or leaf nodes. Based on the inference made from the current decision node, a child node is selected for further branching. DT training involves the recursive partitioning of the training data, which is split into increasingly homogeneous subsets based on a splitting criterion. There are several commonly used criteria to choose from, such as Information Gain and Gini Index [10, 11].

2.2.3.4 Artificial Neural Networks

An Artificial Neural Network (ANN) is a supervised machine learning method that is widely used for pattern classification and related problems [11, 12]. ANNs consist of neurons and weighted connections between neurons. Neurons are arranged in layers, and weighted connections link the neurons in different layers. A value is associated with each connection, referred to as weight, corresponding to the synaptic strength of neuron connections. The behavior of an ANN depends on both the weights and the input–output function, referred to as transfer function. This function typically falls into one of three categories, namely, linear, step, and sigmoid. The back-propagation algorithm is widely used for training ANNs [11]; it minimizes the difference between actual outputs of ANNs and the desired values using gradient descent. When the difference is less than a pre-defined threshold, referred to as the performance goal, the training is deemed to be complete.

2.2.3.5 Recurrent Neural Networks

A recurrent neural network (RNN) is a class of deep-learning artificial neural networks where connections between units form a directed cycle [13]. This creates an internal state of the network, which allows it to exhibit dynamic temporal behavior. Unlike traditional feed-forward neural networks, the hidden-layer nodes of a recurrent neural network maintain an internal state (memory) that is updated with new inputs fed to the network. Those nodes make decisions based on both the current input and what has come before. Recurrent neural networks can make use of the internal state to process relevant data in arbitrary sequences of inputs, such as time series. RNNs can thus be used for anomaly detection in time series [14]. In addition, the hidden state in RNNs is shared over time and thus can contain information from an arbitrarily long window.

As shown in Fig. 2.7, assume that x_t, o_t, and S_t represent the input, output, and state, respectively, of an RNN at time t. We can see that S_t not only depends on the input x_t, but also on all its previous states. If we unfold such a recurrence in the flow graph, we can see that S_t is a function of S_{t-1} and x_t with corresponding weight W and bias b:

$$S_t = f(S_{t-1}, x_t, W, b) \tag{2.1}$$

Note that the weight matrix W can represent any weight between two nodes, and the bias vector b allows values to be adjusted additively. More specifically, $W^{(SS)}$

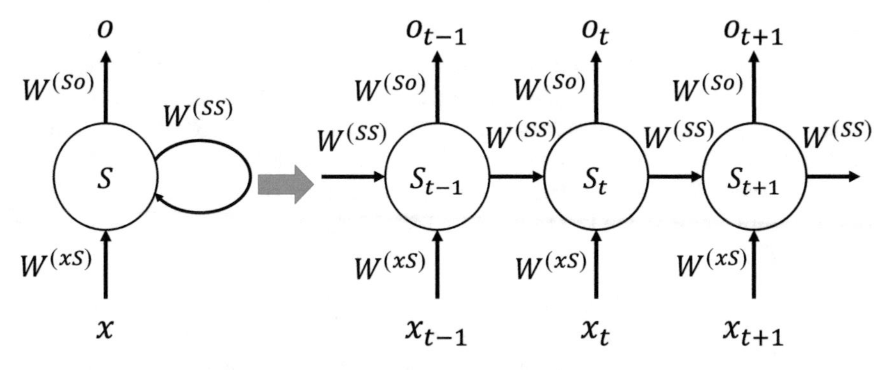

Fig. 2.7 The flow of recurrent neural network (RNN)

represents the weight between current state S_t and previous state S_{t-1}, $W^{(xS)}$ represents the weight between input x_t and current state S_t, and $W^{(So)}$ represents the weight between current state S_t and output o_t. Note that $W^{(SS)}$ is used over successive computations of system state S. To learn arbitrarily complex time series, the function $f(\cdot)$ should provide a non-linear transformation. For example, the hyperbolic tangent *tanh* function is used here to provide non-linearity in the range $(-1, 1)$. Each node in the network can now be formulated as:

$$\begin{aligned} S_t &= tanh(W^{(SS)}S_{t-1} + W^{(xS)}x_t + b_S) \\ o_t &= W^{(So)}S_t + b_o \end{aligned} \tag{2.2}$$

To evaluate the performance of the RNN system, the output, o, needs to be compared to a 'target', y, through a loss function, L. The mean squared error is a common choice for the loss function, as shown below:

$$L(o, y) = \frac{1}{T}\sum_t (o_t - y_t)^2 \tag{2.3}$$

where T is the length of the time series. The objective of training the RNN system is now mapped to finding optimal weight matrices that can minimize the loss function. The stochastic gradient descent (SGD) approach is widely used here to achieve this optimization goal without high computational cost [15]. In SGD, the weight parameters are updated with a learning rate α until a convergence criterion is met:

$$\delta W = -\alpha \frac{1}{|M|} \sum_{(x_m, y_m)\in M} \frac{\partial L(o, y_m)}{\partial W} \tag{2.4}$$

where M is the data points used for training. However, such an SGD procedure can suffer from the vanishing gradient problem, which makes learning long-range dependencies difficult. The vanishing gradient is due to successive multiplications of the derivative of $tanh$ which is bound within (0, 1]. A number of techniques have been developed to overcome the difficulties of learning long-term dependencies. The Long Short Term Memory (LSTM) is widely used among these techniques because it explicitly introduces a memory unit, called the cell, into the network so that long-term historical information can be recalled as needed [16]. Besides the input x_t and the system state S_t, a new type of element called memory cell C_t is introduced. This new memory element can keep or update information over many time steps and it consists of:

1. What information will be discarded from the cell state. This decision is made by a sigmoid function σ called the forget gate layer $f_t = \sigma(W^{(f)}(S_{t-1}, C_{t-1}, x_t) + b_f)$. The layer f_t outputs a number between 0 and 1, where 1 (0) represents "completely keep" ("completely forget").
2. Decide what new information will be stored in the cell state. The decision consists of two parts: a sigmoid function σ called the input gate layer i_t, decides which values will be updated, and a $tanh$ function g_t that creates a vector of new candidate values: $i_t = \sigma(W^{(i)}(S_{t-1}, C_{t-1}, x_t) + b_i)$; $g_t = tanh(W^{(C)}(S_{t-1}, x_t) + b_C)$. Note that $C_t = f_t \times C_{t-1} + i_t \times g_t$.
3. Decide what will be output from the system. A sigmoid function σ is used to decide what parts of the cell state will be output and a $tanh$ function is used to update the system state S_t: $o_t = \sigma(W^{(o)}(S_{t-1}, C_{t-1}, x_t) + b_o)$; $S_t = o_t \times tanh(C_t)$.

LSTM can keep information for a long time because it can propagate information without successive multiplication of fractions. The information can be retained indefinitely if $f_t = 1$ and $i_t = 0$.

We have implemented the above machine-learning techniques in our anomaly detector, and results for commercial core router systems are presented in Sect. 2.4.

2.2.4 Feature-Categorization-Based Hybrid Anomaly Detection

A single class of anomaly detection methods is effective for only a limited number of time-series types. Therefore, we propose a feature-categorization-based hybrid method whereby each class of features can be classified by the most appropriate anomaly detection method. The key idea here is that the statistical characteristics of a feature summarizes its normal behavior and anomalies are considered as statistical outliers. Therefore, if some features have similar statistical characteristics, an anomaly detector that detects anomalies in one feature will also be effective for the other features. Figure 2.8 illustrates the proposed feature-categorization-based hybrid anomaly detection. First, time-series data of different features extracted from the core router system is fed to a KPI-category identification component. Since features belonging to different KPI categories often exhibit significantly different statistical characteristics across the timeline, natural language processing techniques are utilized here to ensure that different KPI categories such as configuration, traffic,

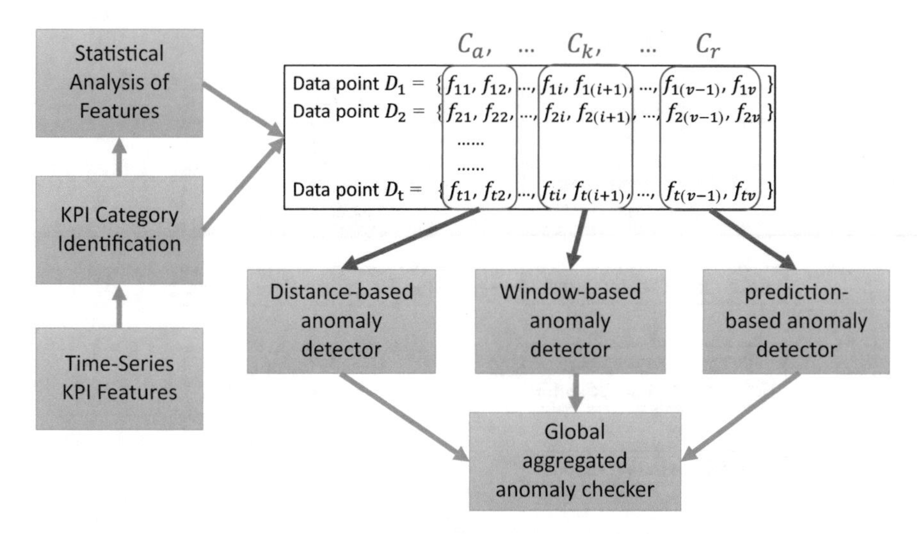

Fig. 2.8 A depiction of feature-categorization-based hybrid anomaly detection

resource type, and hardware can be identified. However, it is also possible that features belonging to different KPI categories have similar trend or distribution across time intervals; therefore, a statistical analysis component is needed to ensure that all features that exhibit similar statistical characteristics are placed in the same class. After these steps, a data point D_t with v features can be divided into different groups $C_a, C_b, \ldots, C_k, \ldots, C_r$, where each group has different statistical characteristics. Next, each group of features is fed to the anomaly detector that is most suitable for this type of features. Finally, the results provided by different anomaly detectors are aggregated so that we can detect an anomaly in terms of the entire feature space.

Initially, the assignment of feature groups to the anomaly detector was hard-coded because the statistical properties of features were assumed to be static. However, in real cases, the statistical properties of features can change as time proceeds. Therefore, an improved voting mechanism is needed so that features can be dynamically assigned to different types of anomaly detectors based on changes in their statistical properties. The flow of the proposed dynamic voting mechanism is shown in Fig. 2.9. We can see that it consists of both off-line and on-line phases. First, an initial set of KPI features are extracted from training data. Their statistical properties are then analyzed and stored. Based on these properties, a set of weights are calculated for each type of anomaly detector. Finally, the initial results are obtained via aggregating results of voting. When new on-line data arrives, we first check whether new features can be extracted from the new data and we update the feature space accordingly. Next, a sliding-window-based statistical analysis component is used to incorporate the new data into the original dataset and analyze their new statistical properties. If the statistical characteristics change significantly, the set of weights for each anomaly detector will also be updated. Finally, the new

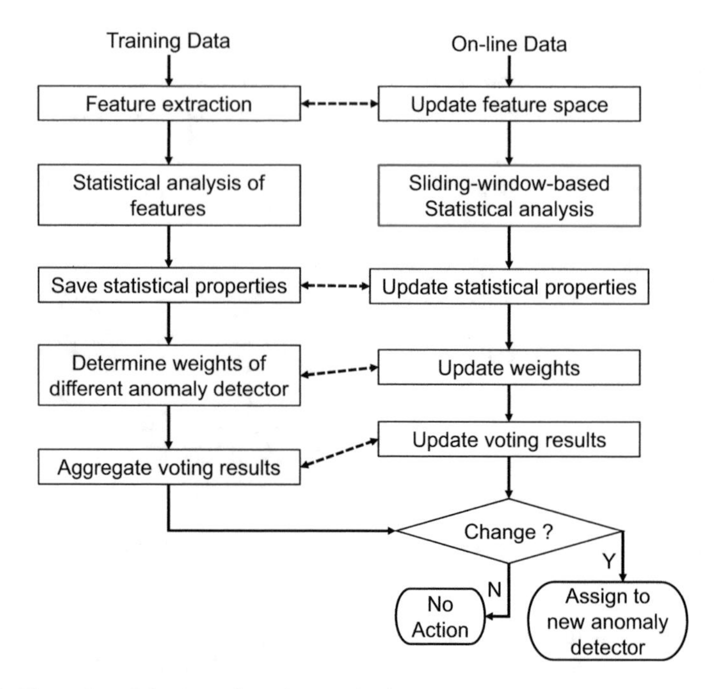

Fig. 2.9 Illustration of the dynamic voting mechanism

voting results are compared with the initial voting results to see whether the current assignment of anomaly detectors needs to be adjusted.

2.3 Correlation-Based Feature Selection

In Sect. 2.2, we have discussed how to apply a number of time-series-based anomaly detection methods to a complex core router system. A key assumption in these methods is that all functions/features of a core network device are treated equally and independently. However, this assumption does not hold true in a real case and can affect the performance of the anomaly detector in several ways [17]. First, since each feature is treated independently, the original anomaly detector will identify a test instance as being anomalous as long as it detect anomalies in any one of features in that test instance. However, it is possible that each feature lies within a normal data region, but the entire instance is still anomalous due to a abnormal combination of these features. Second, since correlation is not considered to remove redundant features, the original anomaly detector will need to monitor and analyze all features in real time. However, the number of feature dimensions will increase from hundreds to tens of thousands when more new features are identified

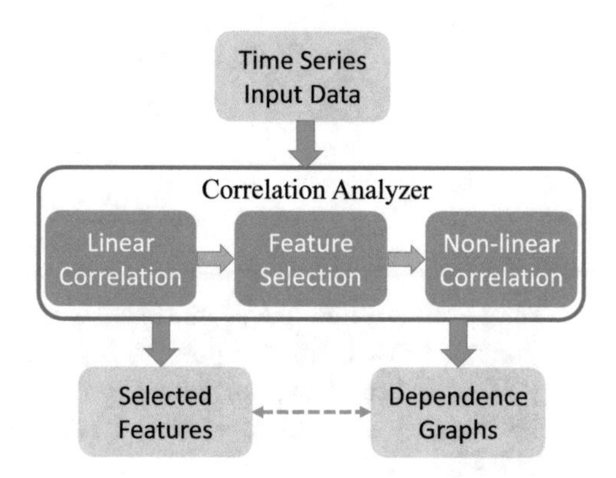

Fig. 2.10 Overview architecture of the proposed Correlation Analyzer

and extracted from raw log data, making it more difficult and time-consuming to detect anomalies. Third, since each feature is treated equally, when anomalies are detected in multiple features, it is difficult to identify the root cause that led to these anomalies. Therefore, in this section, we focus on exploring correlation/dependence among features in order to remove irrelevant and redundant features before applying the proposed anomaly detector.

Figure 2.10 presents an outline of the proposed correlation analyzer. The input time-series data is fed to the correlation analyzer. It then goes through three components: the linear correlation component, the feature selection component, and the non-linear correlation component in sequential order. Finally, the correlation analyzer outputs a number of correlated feature groups. An effective feature subset can be generated by selecting most representative features from these correlated groups. Furthermore, different types of relationships among features within each group can be represented by a dependence graph $G = (V, E)$, where the set of vertices V represent feature candidates and the set of edges E represent dependent relationships between features. Therefore, a dependence graph is generated for each group of features.

2.3.1 Linear Correlation Analysis Between Features

Linear correlation is a widely used statistical relationship between two sets of data and it refers to the extent to which two random variables have a linear relationship with each other. The Pearson product-moment correlation coefficient is a commonly used measure in linear correlation. It is obtained by dividing the covariance of the two variables by the product of their standard deviations. For example, assume that X and Y are two sets of time series. The linear correlation coefficient $Corr(X, Y)$ can be calculated as:

$$\begin{aligned}\rho_{X,Y} &= \mathrm{Corr}(X, Y) \\ &= \frac{\mathrm{cov}(X, Y)}{\sigma_X \sigma_Y} \\ &= \frac{E[(X-\mu_X)(Y-\mu_Y)]}{\sigma_X \sigma_Y}\end{aligned} \tag{2.5}$$

The absolute value $|\rho_{X,Y}|$ must satisfy the inequality $|\rho_{X,Y}| \leq 1$. If $\rho_{X,Y} = 1$, we conclude that there is a positive linear relationship between X and Y, while $\rho_{X,Y} = -1$ indicates a negative linear relationship between X and Y. The closer the coefficient $\rho_{X,Y}$ is to either -1 or 1, the stronger the linear correlation between the variables X and Y.

In our correlation analyzer, input data is first fed to the linear correlation component. In this component, Pearson correlation coefficients are calculated for each pair of features and then compared with a predefined threshold. Feature pairs whose coefficient's absolute value is larger than the threshold are selected as linear-dependent feature pairs. The remaining features are fed to the feature selection that is described next and the non-linear correlation components.

2.3.2 *Feature Selection*

Feature selection, also referred as subset selection, is used to select an effective, but reduced, set of features [18]. Non-linear-dependent features obtained from the previous linear correlation component are fed to our feature-selection component. Therefore, the goal of the feature-selection component is to identify a set of most important features for each non-linear-dependent feature.

One of the most popular solutions for the subset-selection problem is based on the metric of minimum-redundancy-maximum-relevance ($mRMR$) [18]. Suppose we have a set of non-linear-dependent features $\mathbf{F} = \{F_1, F_2, \ldots, F_N\}$. For each feature F_i, we treat it as a target feature A_i and the remaining $N - 1$ features $\{F_1, \ldots, F_{i-1}, F_{i+1}, \ldots, F_N\}$ as its feature candidate set T_i. For each target feature A_i and one of its feature candidate T_{ij}, their mutual information $I(A_i, T_{ij})$ is calculated as shown in (2.6):

$$I(A_i, T_{ij}) = \int\int p(A_i, T_{ij}) \log \frac{p(A_i, T_{ij})}{p(A_i)p(T_{ij})} dA_i dT_{ij} \tag{2.6}$$

where $p(A_i)$ and $p(T_{ij})$ are probability density function of A_i and T_{ij}, and $p(A_i, T_{ij})$ is the joint probability density function of A_i and T_{ij}. We then calculate the relevance value $D(A_i, T_i)$ between the feature candidate set T_i and target feature A_i as follows:

$$D(A_i, T_i) = \frac{1}{|T_i|} \sum_{T_{ij} \in T_i} I(A_i, T_{ij}) \tag{2.7}$$

The MaxRel set $T_i' = \{T_{i1}', T_{i2}', \ldots, T_{im}'\}$ is a selected subset of the top m feature candidates having the highest relevance value. The value of m is determined by counting the number of features whose relevance value is larger than a predefined threshold of low relevance. The set T_i' is further evaluated by computing its redundancy value $R(T_i')$ as shown below:

$$R(T_i') = \frac{1}{|T_i'|^2} \sum_{T_{ij}' \in T_i'} \sum_{T_{ik}' \in T_i'} I(T_{ij}', T_{ik}') \tag{2.8}$$

where $I(T_{ij}', T_{ik}')$, $T_{ij}' \neq T_{ik}'$, is the mutual information between T_{ij}' and T_{ik}'. We then calculate the minimum-redundancy maximum-relevance ($mRMR$) value as follows:

$$mRMR(T_i') = D(A_i, T_i') - R(T_i'). \tag{2.9}$$

We can next determine the minimum-redundancy-maximum-relevance ($mRMR$) feature candidate subset T_i^* for target feature A_i with the largest $mRMR$ value, as follows:

$$T_i^* = \max_{T_i'}\{mRMR(T_i')\}. \tag{2.10}$$

A simple example of $mRMR$-based feature selection is shown in Fig. 2.11. For a target feature A_1, its original set of feature candidates is $T_1 = \{T_{11}, T_{12}, T_{13}, T_{14}, T_{15}\}$. After applying the $mRMR$-based feature selection, the low-relevance features T_{11} and T_{13}, and the highly redundant feature T_{14} are removed, and the final $mRMR$ feature candidate subset for A_1 is $T_1^* = \{T_{12}, T_{15}\}$.

Therefore, for each non-linear-dependent feature F_i, we can obtain a $mRMR$ subset T_i^* containing feature candidates that are most likely to have strong non-linear relationships with F_i. Such feature-correlation group (F_i, T_i^*) is then fed to the following non-linear correlation component for fine-grained dependence analysis.

T_{11}	T_{12}	T_{13}	T_{14}	T_{15}	A_1
0.8	0.2	0.1	0.2	0.3	0.35
0.9	0.1	0.2	0.1	0.1	0.05
0.7	0.1	0.3	0.1	0.2	0.14
0.8	0	0.4	0.1	0.3	0.27

T_{12}	T_{15}	A_1
0.2	0.3	0.35
0.1	0.1	0.05
0.1	0.2	0.14
0	0.3	0.27

Fig. 2.11 An example of $mRMR$-based feature selection

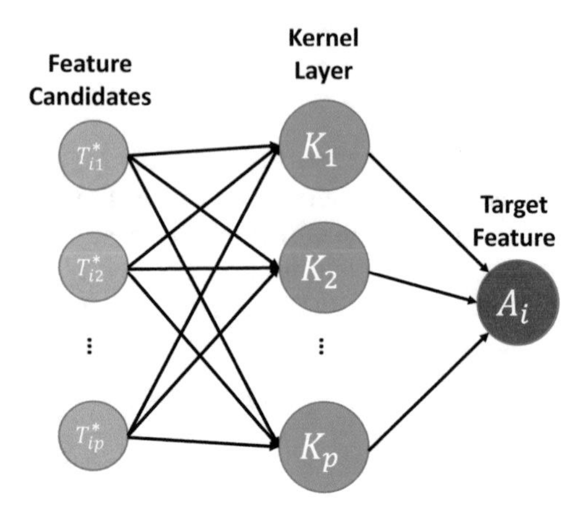

Fig. 2.12 An overview of kernel-based nonlinear correlation between features

2.3.3 Non-linear Correlation Analysis Between Features

Machine-learning techniques are widely used to classify instances based on their feature set. Among these techniques, kernel methods such as support-vector machine (SVM) are effective in finding non-linear types of relationships in the dataset because they can implicitly map their inputs into higher-dimensional feature spaces, solving nonlinear problem using linear classification rules [11].

Figure 2.12 explains how we apply kernel-based methods to identify nonlinear correlations among features. For each feature correlation group (F_i, T_i^*) received from the feature-selection component, a set of kernel nodes are built from training data to determine which kinds of combinations of feature candidates T_i^* can give better performance in predicting changes in target feature F_i. The feature combination that has highest accuracy will be considered as the final dependence graph for the target feature F_i. If the accuracy is lower than a predefined threshold, F_i will be considered as an independent feature that does not rely on any other features.

The *correlation confidence* is defined in our correlation analyzer to indicate how certain the correlation analyzer is about the detected correlations among features. It is measured as a value between zero and one. The closer the *correlation confidence* is to one, the higher is the certainty of correlation that it indicates. For example, the classic coefficient of determination $R^2 = 1 - RSS/TSS$, where RSS and TSS represent the residual sum of squares and total sum of squares of the data, can be used to indicate how well features are correlated with each other.

2.4 Experimental Results

The commercial core router system used in our experiments consists of a number of different functional units such as the main processing unit, line processing unit, switch fabric unit, etc. A local monitoring agent is deployed in each core router system to periodically send collected data to our centralized anomaly detector server. After offline training is completed, the anomaly detector starts to identify whether any new anomalies have occurred in the monitored core routers. The volume of data needed for offline training depends on both the number of extracted features and the number of running tasks. A total of 602 features are monitored and sampled every 30 min. Two test cases with different lengths of monitoring period are discussed: (1) 15 days of operation (720 time points); (2) 30 days of operation (1440 time points).

2.4.1 Anomaly Insertion

Since the commercial core router is designed to be highly reliable, the occurrences of anomalies are rare events during system operation. We therefore inserted different types of synthetic anomalies into our system so that the effectiveness of our proposed methods can be fully validated. Three types of anomalies were considered during our experiments [1, 2]:

(1) Point anomaly: An individual observation point has much higher or lower values than other point. The occurrences of this kind of anomalies are independent of each others. Figure 2.13 shows an example of inserting point anomalies into a traditional Gaussian Process. We can see that after inserting point anomalies, several time points have much higher data values than the normal data range.
(2) Collective anomaly: The behavior of a collection of observation points deviates from normal conditions. The individual points within a collective anomaly are not anomalous by themselves. An example of the insertion of collective anomalies into a bursty distribution is shown in Fig. 2.14. The normal behavior of a bursty distribution is that bursty values appear and disappear within a short period of time while the abnormal behavior of collective anomalies is that several bursty values last for an abnormally period of time.
(3) Trend anomaly: The behavior of an individual observation point violates the overall trend of the entire time series, making it anomalous within a context. Figure 2.15 shows an example of inserting trend anomalies into a monotonic-function-based process. The inserted trend anomalies do not lie outside of normal data range, but they disrupt the increasing trend of the process.

2.4.2 Feature Selection and Categorization

Since in our experiments, the number of extracted features (602) is similar to the number of sample points (720), directly monitoring and detecting anomalies is time-consuming and ineffective. Therefore, it is essential to first utilize the correlation analyzer to select important features.

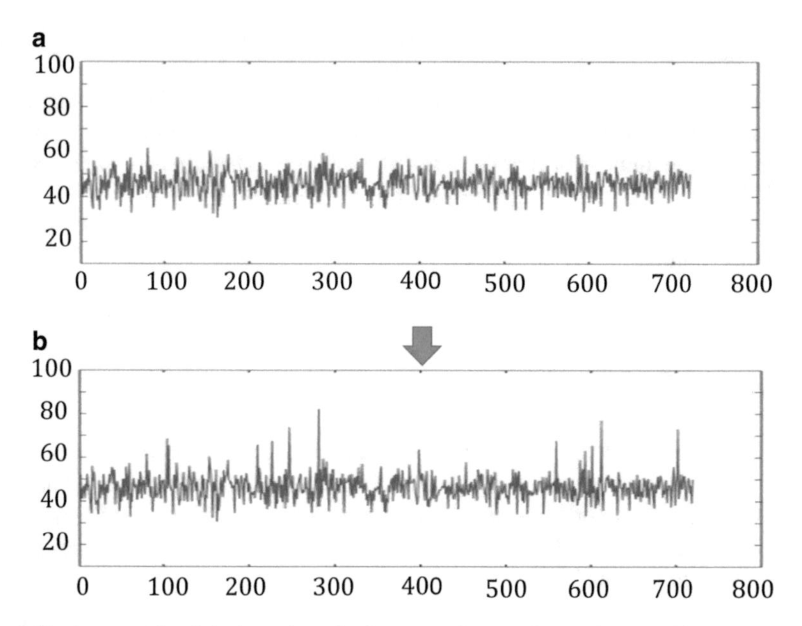

Fig. 2.13 An example of the insertion of point anomalies. (**a**) Data without point anomalies. (**b**) Data with point anomalies

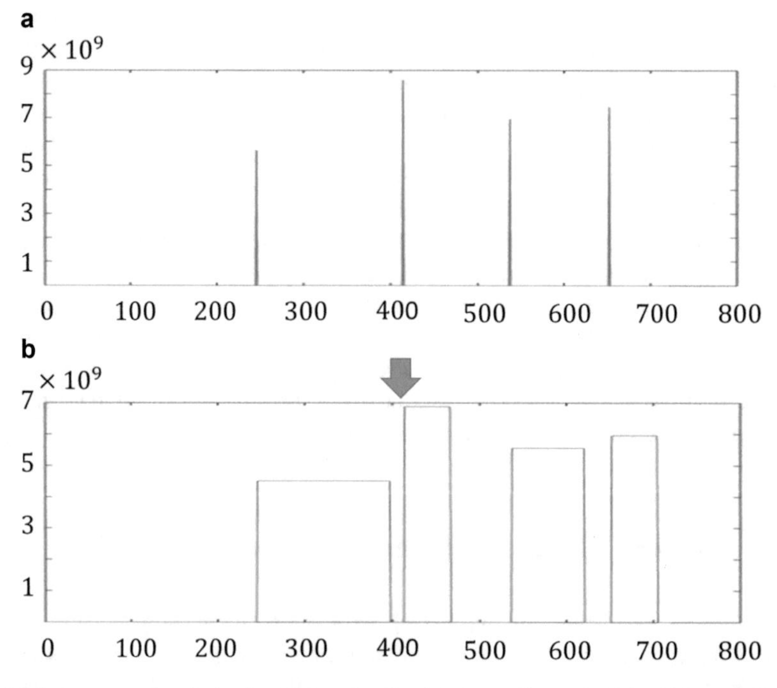

Fig. 2.14 An example of the insertion of collective anomalies. (**a**) Data without collective anomalies. (**b**) Data with collective anomalies

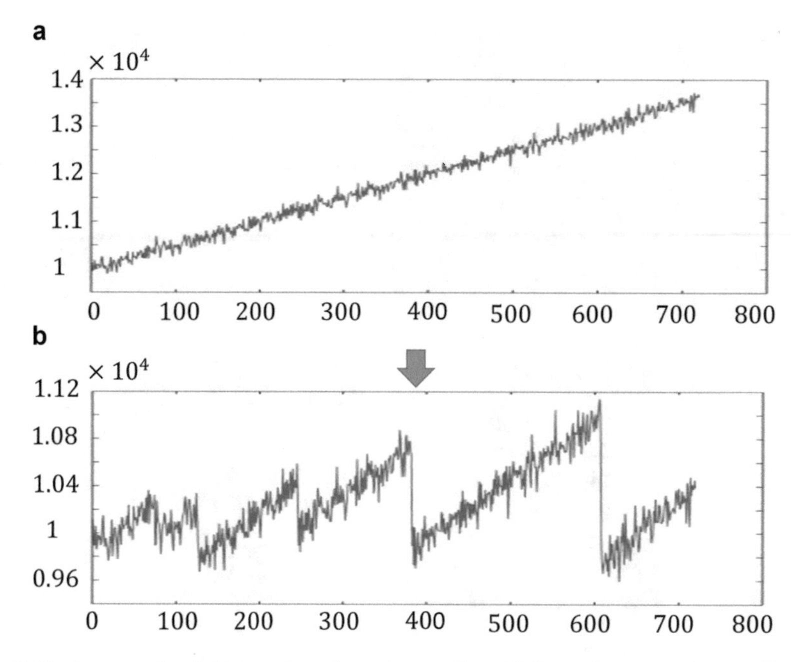

Fig. 2.15 An example of the insertion of trend anomalies. (**a**) Data without trend anomalies. (**b**) Data with trend anomalies

As described in Sect. 2.3, the proposed correlation analyzer determines correlated groups and dependence graphs to guide the selection of the important features. Figure 2.16 shows a sample output of the correlation analyzer. We can see that after feeding the entire feature space to the correlation analyzer, four correlated feature groups are formed: one linear increasing group $\{F_2; F_5\}$, two nonlinear kernel groups $\{F_3, F_4, F_{11}; F_{12}\}$ and $\{F_{15}, F_{18}, F_{20}; F_{25}\}$, and an independent group $\{F_{31}\}$. The correlation confidence within each feature group is strong while the correlation confidence between different feature groups is weak. We find that the correlation analyzer can partition features into disjoint clusters based on their inter-correlation. The results of such correlation-based feature clustering are summarized in Table 2.1. We can see that only 12 out of 602 features are identified as being in independent groups (clusters with a single element), which implies most features are correlated. Moreover, if we choose a single target feature within each cluster to represent this cluster, only 114 features are needed to represent the entire feature space, reducing the number of feature dimensions by 79%. The selected features are then grouped in five categories based on their statistical characteristics, as shown in Table 2.2.

We can see that these 114 features are divided into five categories based on their statistical characteristics, where features within each category have similar statistical characteristics.

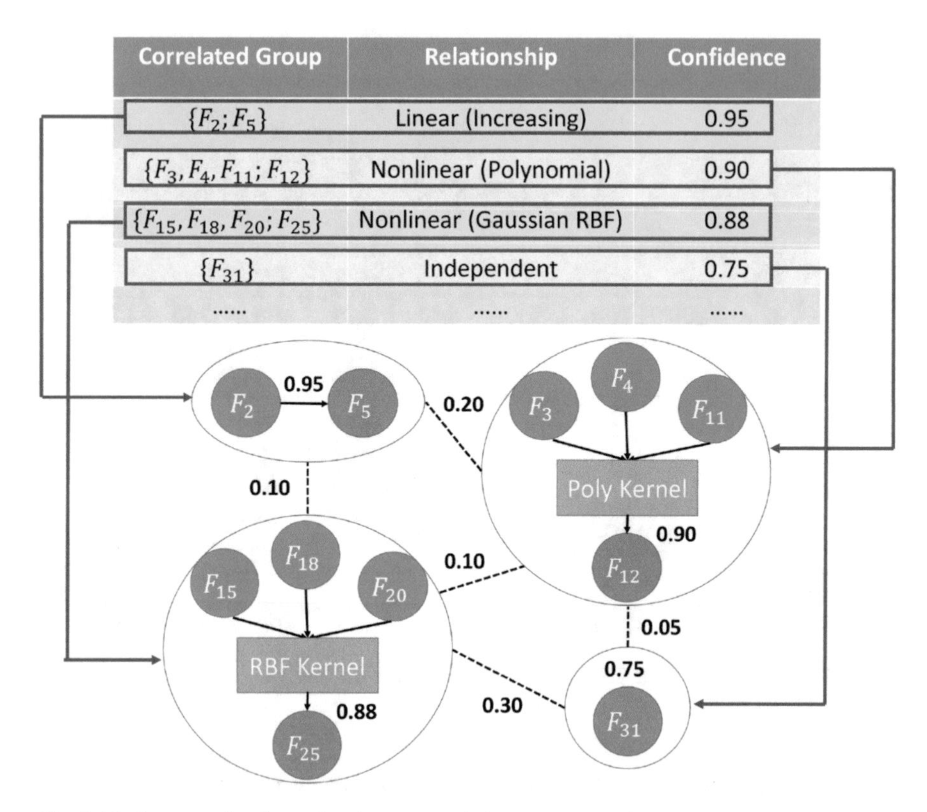

Fig. 2.16 An example of correlated groups and dependence graphs generated by the proposed correlation analyzer

Table 2.1 Results on correlation-based feature clustering

Size of clusters	Number of clusters	Number of features
1	12	12
2	55	110
3	9	27
4	15	60
8	4	32
9	21	189
16	6	96
29	1	29
47	1	47

1. Gaussian Process: Features in this category, e.g., CPU/Memory usage, remain relatively stable across the temporal domain. Therefore, point anomalies occur if some data points are significantly lower or higher than the stable value.
2. Monotonic Function: Features in this category, such as OSPF neighbor uptime, have a monotonically increasing/decreasing trend across the temporal domain.

Table 2.2 Feature categories corresponding to the extracted time-series data for the core router

Feature category	Number of features	Representative features	Potential anomalies
Gaussian process	30	CPU/memory usage	Point anomaly
Monotonic function	14	OSPF neighbor uptime	Point/trend anomaly
Cyclic process	12	ARP learnt count	Point/trend anomaly
Bursty	48	Interface input/output rate	Collective anomaly
Others	10	NP exception/interrupt	Point anomaly

Besides point anomalies, trend anomalies can also occur if some data points together interrupt the monotonic trend.

3. Cyclic Process: Features in this category, such as ARP learnt count, have a recurrent trend across the temporal domain. Trend anomalies occur if the periodicity changes or disappears.
4. Bursty: Features in this category, such as Interface input/output rate, have a set of sudden peak/valley values across the temporal domain. Collective anomalies occur if a peak/valley lasts for an abnormal period.
5. Others: Features in this category are those who do not belong to any one of the above four categories. Therefore, both categorical and continuous features that have irregular patterns are addressed in this category.

2.4.3 Anomaly Detection

To evaluate the performance of different anomaly detection methods, we use a fourfold *cross-validation* method, which randomly partitions the extracted time series dataset into four groups. In each round of experiment, one group is regarded as the test dataset while all the other groups are used for training. Formally, if the total size of the time-series dataset is m, then the size of training and test time-series dataset are $3/4 \times m$ and $1/4 \times m$, respectively. Different types of anomalies are randomly inserted into the test cases. The success ratio (SR), referred to as a percentage, is the ratio of the number of correctly detected anomalies to the total number of anomalies in the testing set. For example, if 10 anomalies are inserted, a SR of 70% means that 7 out of 10 anomalies are correctly detected. In addition to the success ratio, the non-false-alarm ratio (NFAR) is also considered as an evaluation metric. It is defined as the ratio of the number of correctly detected anomalies to the total number of alarms flagged by the anomaly detector.

Two experiments are carried out for evaluation. In the first experiment, we apply different anomaly detection methods to each single-feature category so that we can obtain the best anomaly detector for each category of features. In the second experiment, we apply different anomaly-detection methods to the entire feature space of the log data so that we can verify whether the proposed feature-categorization-based hybrid method is effective.

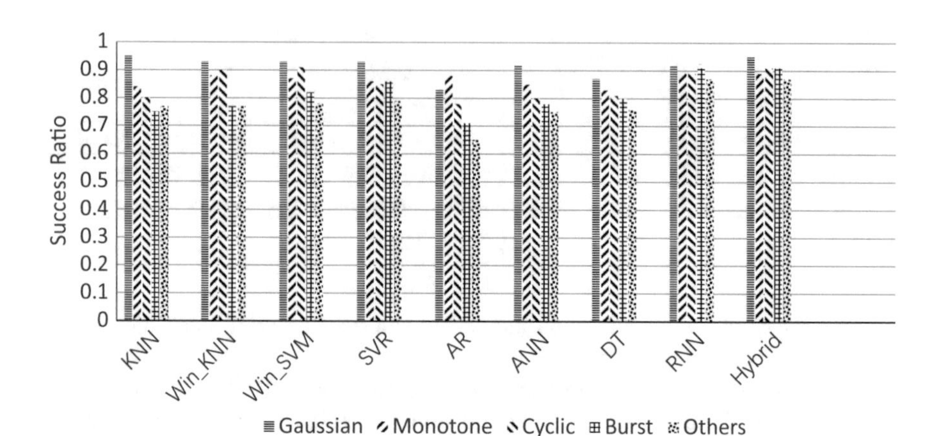

Fig. 2.17 Success ratio of different anomaly-detection methods applied to each feature category (data set 1)

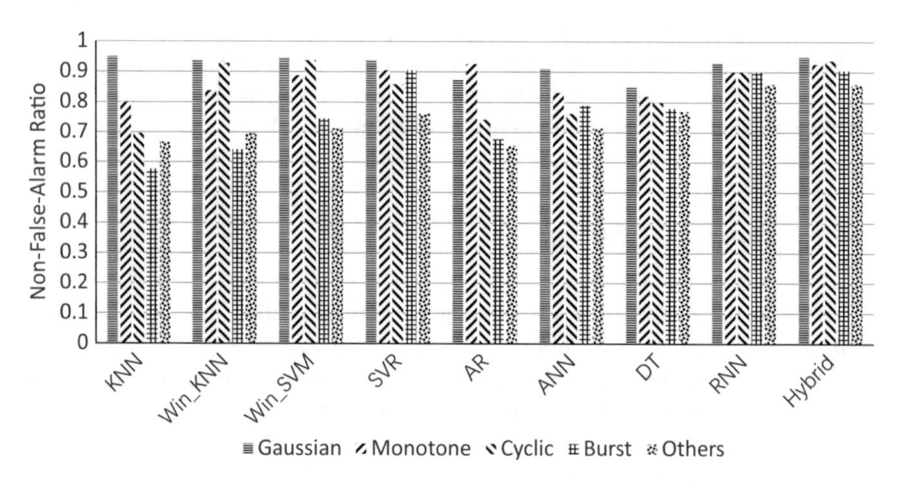

Fig. 2.18 Non-false-alarm ratio of different anomaly-detection methods applied to each feature category (data set 1)

C.I. Results for First Time-Series Data (15 Days of Router Operation)

Figures 2.17 and 2.18 show the success ratio and non-false-alarm ratio of different anomaly detection methods applied to each feature category. The results can be summarized as follows:

1. Distance-based anomaly detection (KNN method): It achieves higher success ratio and non-false-alarm ratio when it is applied to features belonging to the Gaussian Process category. On the other hand, it performs the worst in detecting anomalies in features classified as Bursty. One possible explanation is that the anomalies in the Gaussian process are typically point anomalies, which have much higher or lower values than the normal data points. Such "outliers" can be

easily detected by measuring their distance from normal data points. In contrast, the bursty distribution has a set of peak values, thus it is difficult for KNN methods to distinguish between normal and abnormal data points.

2. Window-based anomaly detection (Window-based KNN and window-based SVM): It performs better than other methods when the features have a cyclic trend because a set of fixed-length overlapping sub-windows can remove the effect of time shifting on the detection of anomalies.
3. Prediction-based anomaly detection (AR, SVR, ANN, DT, RNN): AR performs well when the features have a monotonic trend, but it cannot effectively detect anomalies in features that have a bursty nature. SVR yields satisfactory results for most features, including the burst category. A possible explanation for this observation is that SVR can accurately predict the normal appearance and disappearance of burst values. Therefore, collective anomalies can be detected when some bursts last for an abnormal period of time. DT achieves relatively stable but low SR and NFAR for different feature categories. In this case, it performs worse than many other methods in the Gaussian and Monotone category, but performs relatively well for the other three feature categories. ANN performs better than AR but worse than the SVR. One possible explanation is that the traditional ANN is a feed-forward network architecture with a relatively small number of layers, which limits its ability of learning long-term dependencies within time series. In contrast, RNN achieves relatively stable and high SR and NFAR for all feature categories. Most importantly, it performs best for the Burst and Others categories, where other methods are less effective. One possible explanation for this phenomenon is that the RNN can memorize behaviors in long-term time series via its feedback architecture, and thus learn complex and irregular patterns.

An example of detecting anomalies in a specific feature called "route age" for a real telecom is illustrated in Fig. 2.19. The feature "route age" refers to the uptime of a specific route. In the normal case, as long as the service in this route is up, the "route age" should increase monotonically over time. However, sometimes the core router system will repeatedly remove and add routes due to internal errors, interrupting the monotonically increasing trend of the "route age". As shown in Fig. 2.19, the "route age" in the abnormal case has encountered four sudden drops over time, and the proposed anomaly detector can not only detect the occurrence of such trend anomaly, but also identify the four time points that are most likely to have such sudden decreases.

Next, we determine the success ratio and non-false-alarm ratio of different anomaly detection methods when they are applied to the entire feature space. The results are shown in Figs. 2.20 and 2.21. We can see that for the nine anomaly detection methods, i.e., KNN, Window-based KNN, window-based SVM, SVR, AR, ANN, DT, RNN, and the feature-categorization-based hybrid method, the success ratios are 78.9%, 80.6%, 82.9%, 84.2%, 74.9%, 79.1%, 78.8%, 89.3% and 91.4%, respectively, and the non-false-alarm ratios are 73.1%, 76.3%, 80.7%, 88.1%, 71.6%, 80.9%, 78.5%, 90.2% and 92.1%. The reason that the proposed

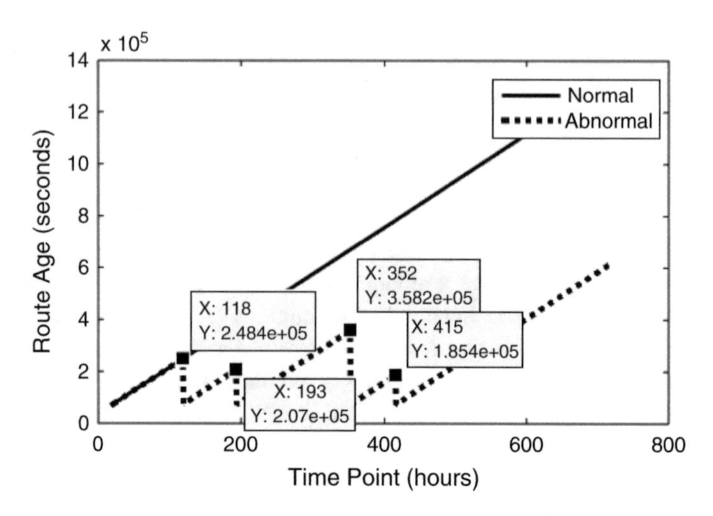

Fig. 2.19 An illustration of detecting anomalies in the feature "route age"

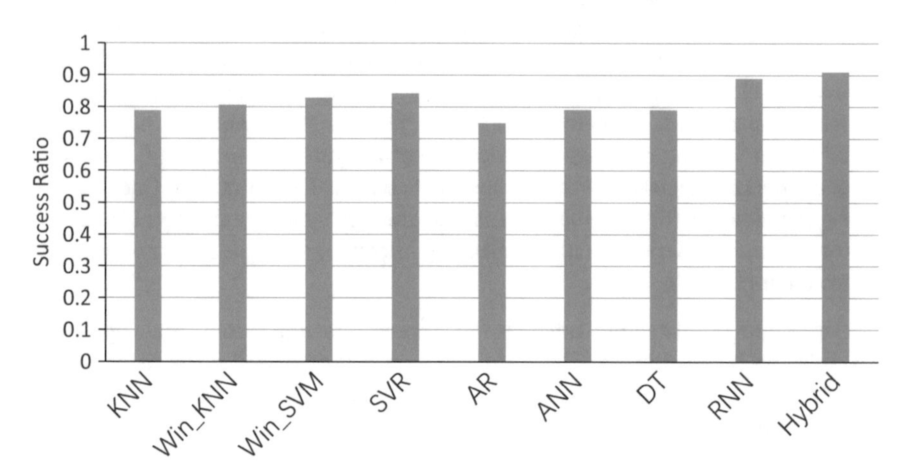

Fig. 2.20 Success ratio of different anomaly-detection methods applied to the entire feature space (data set 1)

feature-categorization-based hybrid method achieves much higher success ratio and non-false-alarm ratio than other methods is that it can overcome the difficulty of adopting a single class of anomaly detection to features with significantly different statistical characteristics. From the receiver operating characteristic (ROC) curve [19] shown in Fig. 2.22, we can also see that the proposed hybrid method is closest to the "perfect classification" point located at the coordinate point (0,1) of the ROC space, which means all anomalies have been detected and no false alarms are generated. In contrast, the AR model is farthest from the "perfect classification" point, which means that it performs worst among all the nine anomaly detection methods in our experiments. All the other anomaly detection methods including

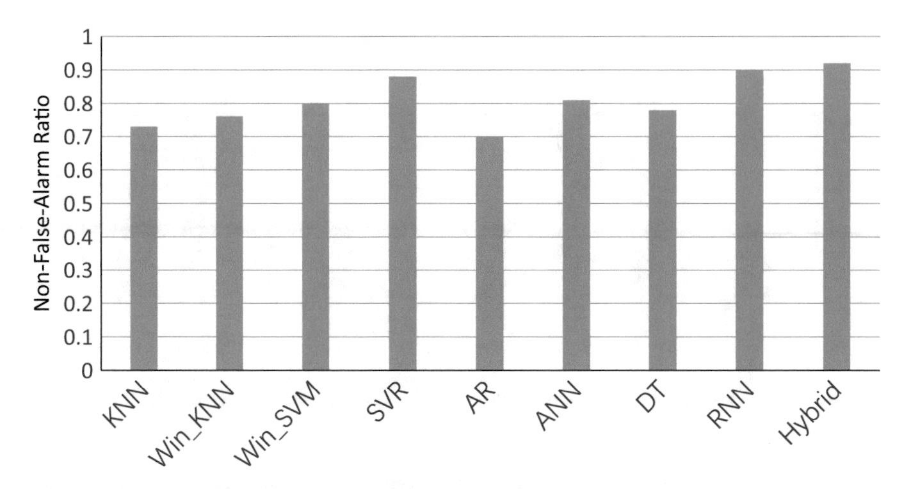

Fig. 2.21 Non-False-Alarm ratio of different anomaly-detection methods applied to the entire feature space (data set 1)

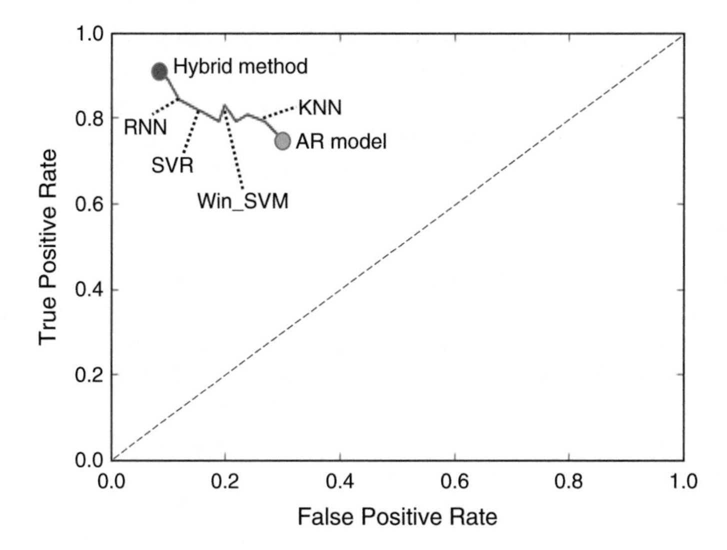

Fig. 2.22 The ROC curve of different anomaly-detection methods applied to the entire feature space

RNN, SVR, window-based SVM, KNN, etc., locate between the hybrid method and the AR model in Fig. 2.22.

The total time cost associated with training and testing for the different anomaly detection methods are also analyzed, as shown in Fig. 2.23. We can see that the distance-based KNN method incurs the least time cost due to its low computational complexity during training. The time cost of SVR, ANN and DT methods are similar, and is much less than the two window-based methods. The RNN approach

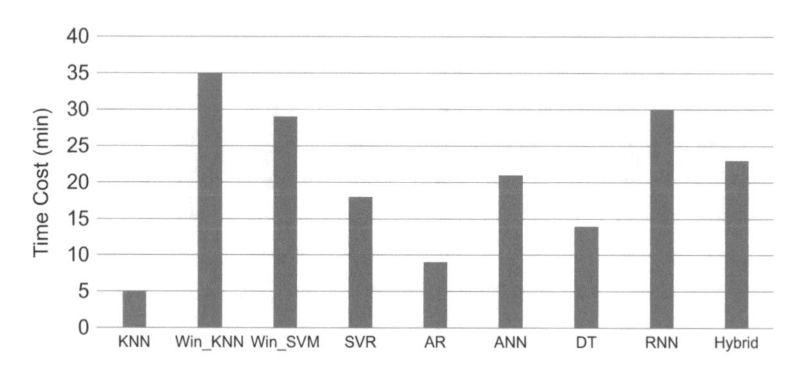

Fig. 2.23 Time cost of different anomaly-detection methods applied to the entire feature space

has higher time cost because it needs to learn a large number of different time dependencies and irregular patterns within time series via a deep-depth neural network. However, it is still comparable to the two window-based methods.

C.II. Results for Second Time-Series Data (1 Month of Router Operation)

Next, we evaluate the performance of different anomaly detection methods on the second dataset, which is extracted from 30 days of core router's operations and consists of 1440 time points. Similar to Figs. 2.17 and 2.18, Figs. 2.24 and 2.25 show the success ratio and non-false-alarm ratio of different anomaly detection methods applied to each feature category. Compared with the result on the first dataset, we can see that:

1. Similar to the results of the first dataset, the distance-based anomaly detection (KNN method) achieves higher success ratio and non-false-alarm ratio for features belonging to the Gaussian Process category. The window-based anomaly detection (Window-based KNN and window-based SVM) performs better when the features are periodical in nature. The prediction-based anomaly detection (AR, SVR, ANN, DT, RNN) can achieve high performance when the statistical properties of features can be predicted accurately.
2. The success ratio and non-false-alarm ratio of most methods decrease significantly in the second dataset. One possible reason is that the number of time points in the second dataset doubles that in the first dataset. Due to the "curse of dimensionality", the longer the time series, the more difficult it is for traditional methods to distinguish between normal and abnormal behaviors.
3. The RNN method achieves high and stable performance for all feature categories in both two datasets. One possible explanation is that RNN can memorize behaviors in long-term time series via its feedback architecture, and thus it is affected less by an increase in temporal dimensionality.

Also, when the nine anomaly detection methods, i.e., KNN, Window-based KNN, window-based SVM, SVR, AR, ANN, DT, RNN, and the feature-categorization-based hybrid method are applied to the entire feature space, the

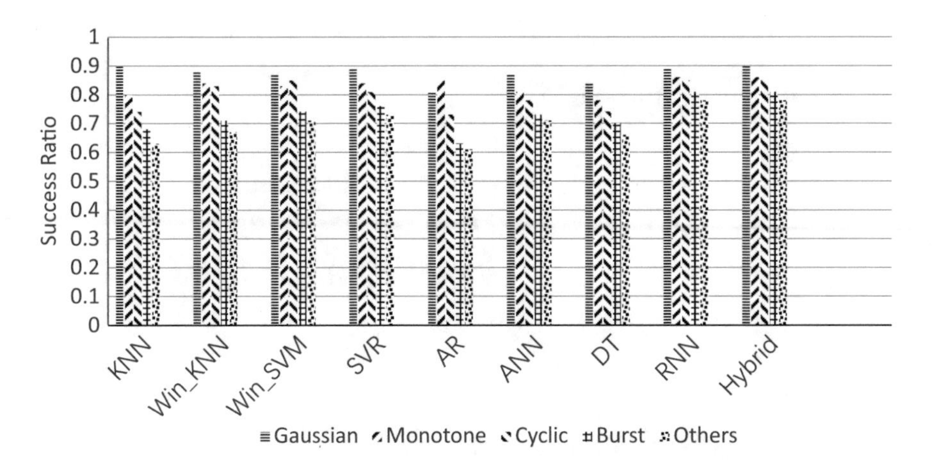

Fig. 2.24 Success ratio of different anomaly-detection methods applied to each feature category (data set 2)

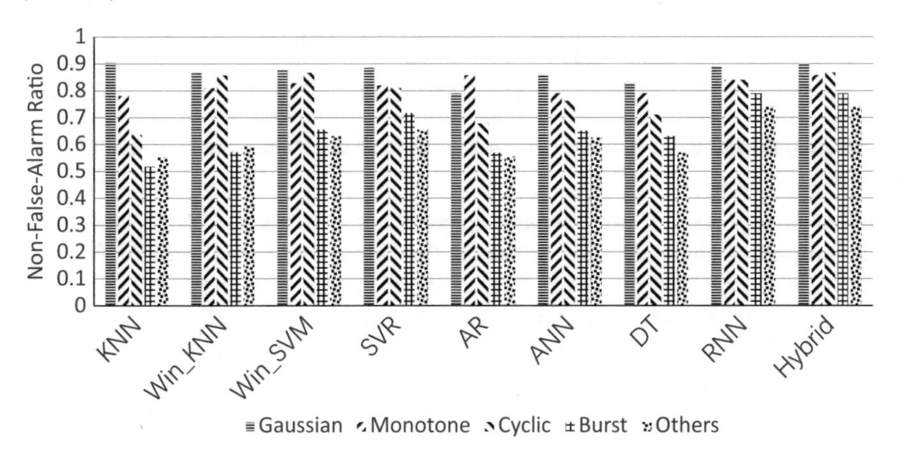

Fig. 2.25 Non-false-alarm ratio of different anomaly-detection methods applied to each feature category (data set 2)

resulting success ratios are 71.6%, 74.3%, 76.1%, 77.5%, 69.6%, 75.1%, 72.2%, 83.3% and 85.4%, respectively, and the non-false-alarm ratios are 61.1%, 66.3%, 68.7%, 73.1%, 63.6%, 71.9%, 63.5%, 80.2% and 82.1%. We can see that although the overall success ratio and non-false-alarm ratio for the second dataset is much lower than that for the first dataset in many cases, the proposed hybrid method still performs relatively well. The reason is that it can dynamically choose the most-appropriate anomaly detectors for different features.

Moreover, since we incorporate a correlation analyzer into the anomaly detector, anomalies caused by a abnormal combination of features can now be successfully detected, as illustrated in Fig. 2.26. We can see that three features F_{i1}, F_{i2}, and F_{i3} lie within the normal data region across the entire temporal domain, but

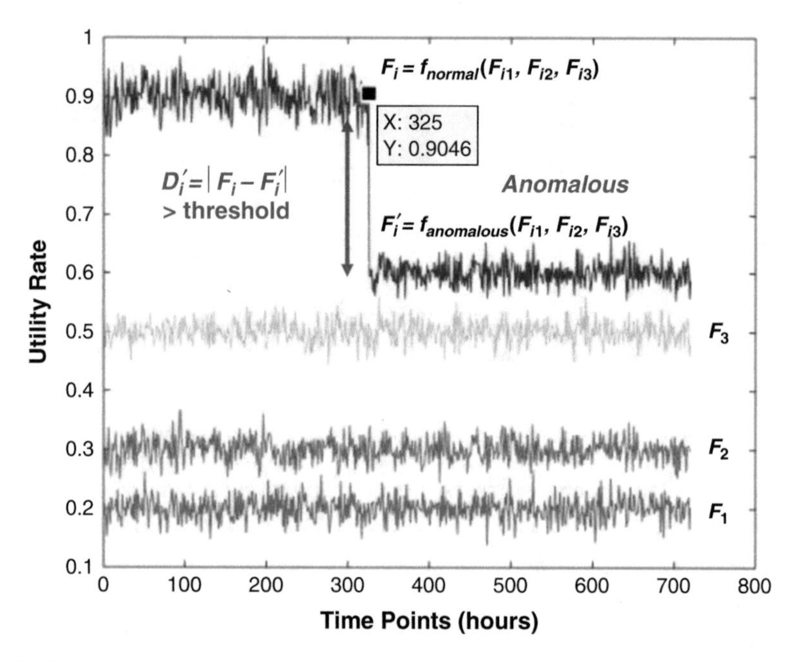

Fig. 2.26 An illustration of detecting anomalies caused by an anomalous combination of features

our anomaly detector still triggers an alarm at time point 325, indicating the appearance of anomalies. The reason is that when data is fed to our correlation analyzer, a correlated group i is formed and the implicit representative feature F_i for this group is a combination of these three features F_{i1}, F_{i2}, and F_{i3}: $F_i = f_{normal}(F_{i1}, F_{i2}, F_{i3})$. When errors occur in the core router system, an anomalous relationship among F_{i1}, F_{i2}, and F_{i3} is formed: $F_i' = f_{anomalous}(F_{i1}, F_{i2}, F_{i3})$. Our anomaly detector can identify such anomalies when the accumulated distance between F_i and F_i' exceeds a predefined threshold.

In summary, in our experiments with a commercial core router system, we have clustered features based on their correlations and selected the most important features to be used in anomaly detection. We have found that the effectiveness of anomaly detection depends on the statistical characteristics of the features. The proposed feature-categorization-based hybrid method incorporates the advantages of different anomaly detection methods, leading to higher success ratios for detecting different types of anomalies.

2.5 Conclusion

We have described the design of a time-series-based anomaly detector for a complex core router system. A correlation analyzer has been implemented to group correlated features and select important features. A number of anomaly-detection techniques have been implemented, among which the feature-categorization-based hybrid method appears to be the most promising. Different types of anomalies have been used to evaluate the effectiveness of the proposed methods.

References

1. A. Patcha, J.-M. Park, An overview of anomaly detection techniques: existing solutions and latest technological trends. Comput. Netw. **51**, 3448–3470 (2007)
2. V. Chandola, A. Banerjee, V. Kumar, Anomaly detection: a survey. ACM Comput. Surv. (CSUR) **41**, 15:1–15:58, (2009)
3. Y. Liao, V.R. Vemuri, Use of k-nearest neighbor classifier for intrusion detection. Comput. Secur. **21**, 439–448 (2002)
4. S. Mukkamala, G. Janoski, A. Sung, Intrusion detection using neural networks and support vector machines, in *Proceedings of the International Joint Conference on Neural Networks*, vol. 2 (2002), pp. 1702–1707
5. J. Herzog, S. Wegerich, K. Gross, F. Bockhorst, MSET modeling of crystal river-3 venturi flow meters, in *Proceedings of the 6th International Conference on Nuclear Engineering* (1998), pp. 1–17
6. K. Vaidyanathan, K. Gross, MSET performance optimization for detection of software aging, in *Proceedings of the ISSRE* (2003)
7. R. Giladi, *Network Processors: Architecture, Programming, and Implementation* (Morgan Kaufmann, Burlington, 2008)
8. V. Vapnik, An overview of statistical learning theory. IEEE Trans. Neural Netw. **10**(5), 988–999 (1999)
9. C. Cortes, V. Vapnik, Support-vector networks. Mach. Learn. **20**, 273–297 (1995)
10. J. Quinlan, Induction of decision trees. Mach. Learn. **1**, 81–106 (1986)
11. F. Ye, Z. Zhang, K. Chakrabarty, X. Gu, Board-level functional fault diagnosis using artificial neural networks, support-vector machines, and weighted-majority voting. IEEE Trans. Comput.-Aided Des. Integr. Circ. Syst. **32**, 723–736 (2013)
12. B. Prieto, J. de Lope, D. Maravall, Reconfigurable hardware implementation of neural networks for humanoid locomotion, in *International Work-Conference on the Interplay Between Natural and Artificial Computation* (2005), pp. 395–404
13. A. Graves et al., Speech recognition with deep recurrent neural networks, in *International Conference on Acoustics, Speech and Signal Processing* (2013), pp. 6645–6649
14. M.S. alDosari, Unsupervised anomaly detection in sequences using long short term memory recurrent neural networks, Master's thesis, 2016
15. L. Bottou, Large-scale machine learning with stochastic gradient descent, in *Proceedings of COMPSTAT* (2010), pp. 177–186
16. S. Hochreiter, J. Schmidhuber, Long short-term memory. Neural Comput. **9**, 1735–1780 (1997)
17. C.-K. Hsu et al., Test data analytics–exploring spatial and test-item correlations in production test data, in *Proceedings of the IEEE International Test Conference (ITC)* (2013), pp. 1–10
18. H. Peng, F. Long, C. Ding, Feature selection based on mutual information: criteria of max-dependency, max-relevance, and min-redundancy, in *IEEE Transactions on Pattern Analysis and Machine Intelligence*, vol. 27 (2005), pp. 1226–1238
19. C.D. Brown, H.T. Davis, Receiver operating characteristics curves and related decision measures: a tutorial. Chemometr. Intell. Lab. Syst. **80**, 24–38 (2006)

Chapter 3
Changepoint-Based Anomaly Detection

In this chapter, we describe the design of a changepoint-based anomaly detector that first detects changepoints from collected time-series data, and then utilizes these changepoints to detect anomalies. Different changepoint detection approaches are implemented to detect various types of changepoints. A clustering method is then developed to identify a wide range of normal/abnormal patterns from changepoint windows. Data collected from a set of commercial core router systems are used to validate the proposed anomaly detector. Experimental results show that our changepoint-based anomaly detector achieves better performance than traditional methods.

The remainder of the chapter is organized as follows. Section 3.1 presents the motivation of changepoint-based anomaly detector for core routers. Section 3.2 describes the framework of changepoint-based anomaly detector. Section 3.3 presents how changepoints are detected and Sects. 3.4 and 3.5 describe how different normal/abnormal patterns are learned using information around changepoints in more detail. Experimental results for a commercial core router system are presented in Sects. 3.6–3.8. Finally, Sect. 3.9 concludes the chapter.

3.1 Motivation

In a complex system such as a core router, data is collected in the form of time series. For example, as shown in Fig. 2.1, a multi-card chassis core router system uses monitors to log features from different functional units [1]. In this work, we improve the way we extract and collect data via taking into account the internal hierarchy of different features. From Fig. 3.1, we can see that features extracted from the global system serve as the first hierarchical level and these features summarize the general conditions of various functional units such as MPU, LPU, SFU, and interface. Next, a set of second-level features are extracted from these functional units to indicate

S. Jin et al., *Anomaly-Detection and Health-Analysis Techniques for Core Router Systems*, https://doi.org/10.1007/978-3-030-33664-6_3

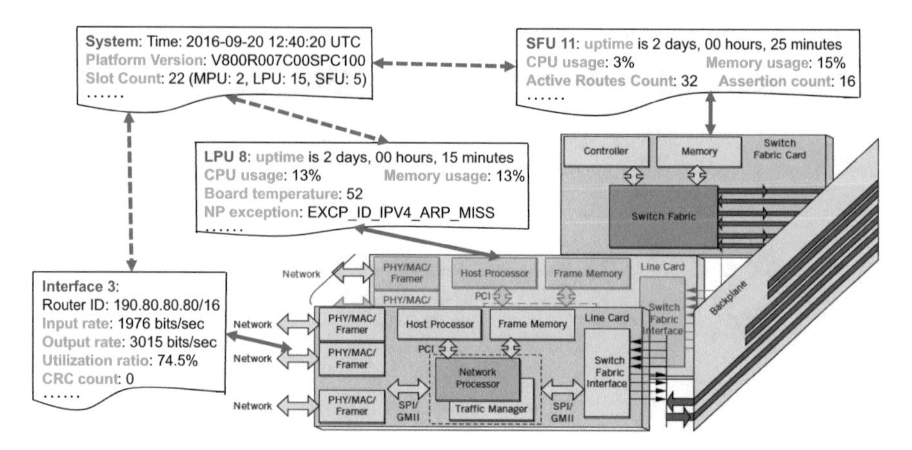

Fig. 3.1 A multi-card chassis core router system [1] and a snapshot of features extracted in a hierarchical way

detailed operation information such as CPU/Memory Usage. Finally, since some components inside these functional units have their own status metrics, several third-level features are also recorded.

As described in Chap. 2, three types of techniques, namely distance-based, window-based, and prediction-based methods have been used to detect anomalies in time-series data [2]. Since a single class of anomaly detection methods is effective for only a limited number of time-series types, a feature-categorization-based hybrid method has also been presented in Chap. 2.

A drawback of the above methods is that they make assumptions that are not valid in realistic scenarios. For example, the supervised prediction-based methods assume that we know labels ("normal" or "abnormal") in advance for the training data. However, raw data collected from commercial core routers is unlabeled. Labeling such raw data requires manual checking by expert technicians, which needs a considerable amount of time and labor. In contrast, the unsupervised distance-based approaches do not require any labels. However, they assume that abnormal instances are rare events that are significantly different from all other historical instances. Such an assumption is not valid for a core router system because some normal instances are also rare. The window-based methods are infeasible for long-term time series instances because their time cost increases significantly with the size of the temporal dimension [2]. The feature-categorization-based hybrid method presented in Chap. 2 makes the assumption that the statistical properties of normal features remain constant across the entire temporal domain. However, new normal patterns can appear as time proceeds and the statistical characteristics of features can change significantly even if no anomalies occur. Therefore, in this work, we address the problem of detecting anomalies in complex core router systems without relying on the unrealistic simplifying assumptions that were made in prior work.

3.2 Framework of Changepoint-Based Anomaly Detection

We propose a two-step changepoint-based anomaly detection scheme, as shown in Fig. 3.2. The key idea is that instead of directly detecting anomalies from a large volume of time-series data, we first detect all the changepoints, which indicate significant scenario changes. The changepoint (CP) windows are built from useful data around each changepoint. Machine-learning algorithms are then applied to these CP windows to determine normal window patterns and anomalous window patterns. When new data arrives, it goes through the changepoint detection procedure, and then the following actions are taken:

1. If no changepoint is detected, the new data will be identified as being normal and no further actions will be taken.
2. As long as any changepoints are detected, a set of new CP windows will be determined and fed to the previously learned normal/abnormal window pattern library. Then for each new CP window, the following actions will be taken:
   (a) If the new CP window is classified as an abnormal window pattern, an anomaly will be reported;
   (b) If the new CP window belongs to the category of normal window patterns, no alert is generated;
   (c) If the new CP window lies outside any existing window patterns, it is identified as a suspect window;
   (d) If any matched rules are found for this suspect window in the expert anomaly rule table, it is labeled as an anomaly and used to update the original abnormal pattern library. Note that the update of the existing pattern library occurs before testing for the next case, the new-coming changepoint window will always be compared with the up-to-date pattern library. However, for

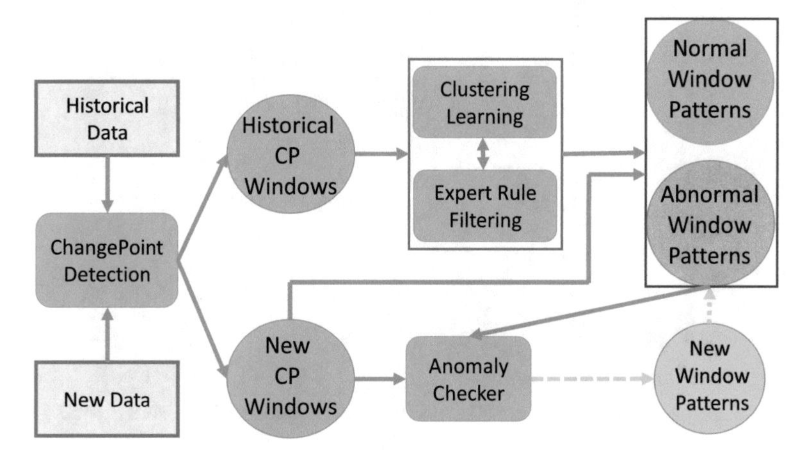

Fig. 3.2 Illustration of changepoint-based anomaly detection

those changepoint windows that had been previously identified using the old pattern library, they will not be tested using this updated pattern library again.

The advantages of the proposed anomaly-detection method are as follows:

1. Only time points around changepoints are considered during model training and testing, significantly reducing the dimensionality of temporal domain.
2. Both abnormal and new/rare normal patterns can be identified, thereby reducing the number of false alarms caused by new/rare normal patterns.

3.3 Changepoint Detection

Changepoints are locations where abrupt changes occur in time-series data. These changes can refer to changes in various statistical properties such as mean, variance, trend, frequency, etc. As shown in Fig. 3.3, a changepoint τ splits a time series into two disjoint segments d_1, d_2 with probability density functions $f(\mu_1, \sigma_1^2)$, $f(\mu_2, \sigma_2^2)$. We can see that the segment d_1 before τ and the segment d_2 after τ have significantly different statistical properties such as mean and variance. Therefore, the problem of changepoint detection can be described as identifying the time points where the statistical properties of the time sequence show heterogeneity or discontinuity.

Assume that we have time-series data $D = \{d_1, d_2, \ldots, d_n\}$, where n is the length of D. A set of changepoints $CP = \{\tau_1, \tau_2, \ldots, \tau_m\}$ partitions D into $m + 1$ disjoint segments $S = \{s_1, s_2, \ldots, s_{m+1}\}$, where m is the number of changepoints. If τ_i

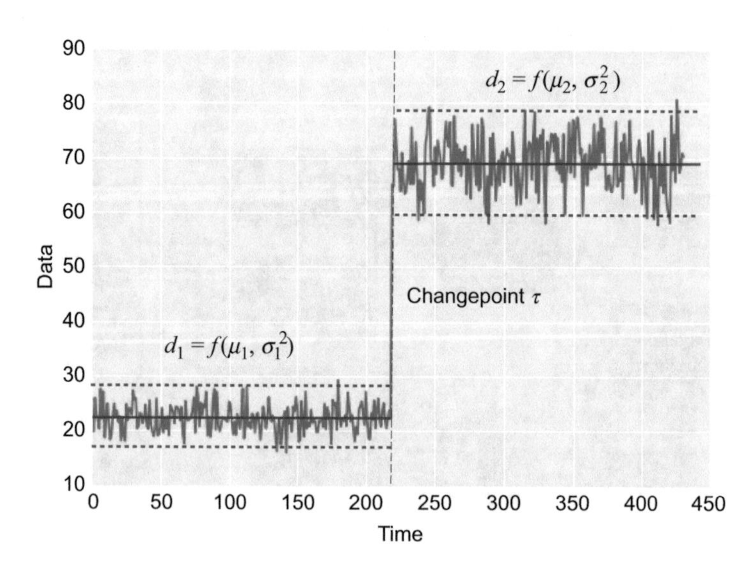

Fig. 3.3 Illustration of changepoint in time series

denotes the time location of the ith changepoint, the value of τ_i will be integers between 1 and $n-1$, $1 \leq \tau_i \leq n-1$, $\tau_i \in \mathbb{Z}$. Assume that the changepoints are ordered, $i < j \iff \tau_i < \tau_j$, where the ith segment s_i contains data between τ_{i-1} and τ_i, $s_i = \{d_{\tau_{i-1}+1}, d_{\tau_{i-1}+2}, \ldots, d_{\tau_i}\}$. Each segment s_i can be represented by a set of statistical properties $\{q_i, \theta_i, \varepsilon_i\}$, where q_i is the statistical model type, θ_i is the set of parameters associated with q_i, and ε_i is a set of external noise parameters [3]. Four classes of changepoint detection approaches, namely density-estimation-based method, density-ratio-estimation-based method, clustering-based method, and Bayesian-based online method are used to detect various types of changepoints. A hybrid multivariate changepoint detection is designed to combine the advantages of these four classes of approaches. These five methods are introduced in the following subsections.

3.3.1 Density-Estimation-Based Method

The maximum likelihood estimation is a widely used technique that can identify these statistical properties given actual observations [4]. As shown in Fig. 3.4, for a time series D with 153 independent and identically distributed observations, its joint probability density function can be specified as $f(d_1, \ldots, d_{153}|\theta) = f(d_1|\theta) \times f(d_2|\theta) \times \ldots \times f(d_{153}|\theta)$. If we take θ as a function variable and all observations $d_1, \ldots, d_{153}$ as fixed parameters, the likelihood function of D can be formed: $L(\theta) = f(d_1, \ldots, d_{153}|\theta) = \prod_{i=1}^{153} f(d_i|\theta)$. Therefore, θ with higher $L(\theta)$ can represent the statistical property of D better. We can see from Fig. 3.4 that if θ is estimated as $N(4, 5^2)$ for all observations, the reconstructed time series is significantly different from D. In contrast, if a set of estimates $\{N(10, 1^2), N(-3, 4^2), N(1, 0^2)\}$ is used for the different segments, the reconstructed time series fits D well. In this

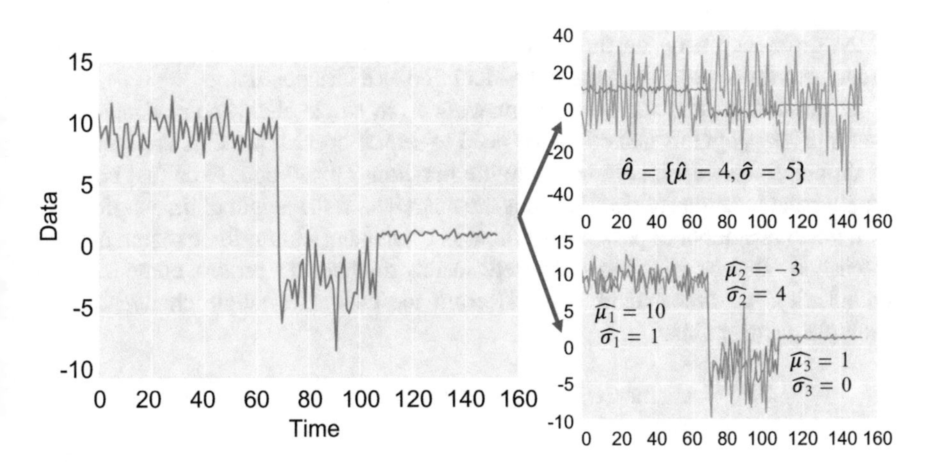

Fig. 3.4 Illustration of maximum likelihood estimation

case, the objective of detecting changepoints is to find an optimal partition that maximizes the product of likelihood functions of all disjoint segments.

To simplify computation, the natural logarithm of the likelihood function, called the log-likelihood, is often used. Therefore, the objective function is the sum of log-likelihood functions of all disjoint segments.

Since detecting a single changepoint is the simplest form of our changepoint detection problem, we describe the maximum-likelihood-based approach for this case. Assume that a single changepoint τ_1 divides a time-series data D into two segments s_1 and s_2 with likelihood functions $p(s_1|\hat{q_1}, \hat{\theta_1}, \hat{\varepsilon_1})$ and $p(s_2|\hat{q_2}, \hat{\theta_2}, \hat{\varepsilon_2})$, respectively, where $p(\cdot)$ represents the probability density function and $\hat{q_i}$, $\hat{\theta_i}$, $\hat{\varepsilon_i}$ are the maximal likelihood estimate of model type, model parameters, and noise factors, respectively. Therefore, the sum of log-likelihood function of s_1 and s_2 can be written as $L(\tau_1) = \log p(s_1|\hat{q_1}, \hat{\theta_1}, \hat{\varepsilon_1}) + \log p(s_2|\hat{q_2}, \hat{\theta_2}, \hat{\varepsilon_2})$. Then the objective of finding the location of τ_1 can be formulated as:

$$\begin{aligned}\tau_1^* &= \underset{\tau_1=1,\ldots,n}{\arg\max}\, L(\tau_1) \\ &= \underset{\tau_1=1,\ldots,n}{\arg\max}(\log p(s_1|\hat{q_1}, \hat{\theta_1}, \hat{\varepsilon_1}) + \log p(s_2|\hat{q_2}, \hat{\theta_2}, \hat{\varepsilon_2}))\end{aligned} \tag{3.1}$$

However, finding τ_1^* does not ensure the existence of a changepoint in current time-series data. We need to check whether the computed log-likelihood is significantly different from the log-likelihood estimated with no changepoints. If we consider D as a single entity, and denote the log-likelihood function with no changepoints as $L(D) = \log p(D|\hat{q}, \hat{\theta}, \hat{\varepsilon})$, then the difference between these two log-likelihood function can be written as:

$$\Delta L = f(L(\tau) - L(D)) \tag{3.2}$$

where $f(\cdot)$ is a transfer function that defines the different types of differences. Only if the difference between these two log-likelihood function ΔL is larger than a predefined threshold c, will the τ_1 be identified as a changepoint.

In order to extend the above framework from single changepoint detection to multiple changepoints detection, we need to search over all possible combinations of different changepoints to compute the maximum likelihood. Note that both the number and locations of changepoints are unknown. If the length of time-series data is n, the total number of possible solutions is 2^n, making a brute-force search method infeasible. Before applying advanced search methods to reduce computational complexity, we first extend Eq. (3.1) from the case of a single changepoint to multiple changepoints:

$$\begin{aligned}CP^* &= \underset{\tau_i=1,\ldots,n}{\arg\max}(\alpha L(CP) - \beta P(m)) \\ &= \underset{\tau_i=1,\ldots,n}{\arg\max}(\alpha \sum_{i=1}^{m+1} (\log p(s_i|\hat{q_i}, \hat{\theta_i}, \hat{\varepsilon_i})) - \beta P(m))\end{aligned} \tag{3.3}$$

In Eq. (3.3), CP^* represents an optimal (i.e., maximum likelihood estimate) combination of changepoints, including both the number of changepoints m and the locations of changepoints $\{\tau_1, \ldots, \tau_m\}$. The parameter $\alpha L(CP)$ is the weighted sum of the log-likelihood of $m+1$ segments split by m changepoints, and $\beta P(m)$ is the weighted penalty function associated with the number of changepoints m to avoid overfitting.

Several heuristic algorithms can be utilized to reduce the search space for the multiple changepoint detection problem. Among these techniques, we consider binary segmentation and the Pruned Exact Linear Time (PELT) algorithm due to their relatively low computational complexity [5, 6]. The basic idea of binary segmentation is divide-and-conquer. It first applies the single changepoint detection procedure on the entire time series. If a single changepoint is identified, the entire time series is divided into two segments. This procedure is repeated on all newly-generated segments until no new changepoints can be found. Binary segmentation reduces the computational complexity from $O(2^n)$ to $O(n \log n)$. The PELT algorithm applies dynamic programming to reduce the search space. Starting from the last changepoint τ_m, the entire time series D is divided into a subset D' before τ_m and a segment s_{m+1} after τ_m. Therefore, the objective function $\max_{\tau_i=1,\ldots,n} \alpha L(CP) - \beta P(m)$ now becomes: $\max_{\tau_i=1,\ldots,n}(\max_{\tau_i=1,\ldots,\tau_m}(\alpha L(CP') - \beta P(m-1)) + L(\tau_m))$. This procedure is executed recursively until we reach the first changepoint τ_1; the optimum CP^* is recorded during this process. A pruning process, which removes solutions that are far from optimality, is also executed at each iteration to further speed up computation, yielding $O(n)$ computational complexity.

In Fig. 3.5, we show how the maximum-likelihood-based approach is used to detect multiple changepoints in our synthetic data. The synthetic time series consists of four different segments, namely s_1, s_2, s_3, and s_4. All segments obey the Gaussian distribution, but with different parameters (μ, σ^2) and time length l. As shown in Fig. 3.5, $s_1 \sim N(3, 0.1)$ with $l_1 = 164$, $s_2 \sim N(-0.9, 0.5)$ with $l_2 = 95$, $s_3 \sim N(0.9, 6)$ with $l_3 = 182$, and $s_4 \sim N(12, 1)$ with $l_4 = 101$. We can see that s_1 and s_2 have similar variance but different means, s_2 and s_3 have similar mean but different variance, and s_3 and s_4 are significantly different in both mean and variance. If the maximum-likelihood-based binary segmentation method is applied to the synthetic time series, the computation can be described as below:

1. Suppose that each time point t_i can divide the synthetic time series into two segments, namely s_a with μ_a and σ_a, and s_b with μ_b and σ_b. The log-likelihood of t_i being identified as changepoint is then calculated as $L(t_i) = \log p(s_a|N(\mu_a, \sigma_a^2)) + \log p(s_b)|N(\mu_b, \sigma_b^2)$. A set of log-likelihood for all time points is therefore obtained $L = \{L(t_1), \ldots, L(t_{441}), \ldots, L(t_{542})\} = \{-2.09, \ldots, -0.105, \ldots, -1.96\}$. The time point t_{441} with maximum log-likelihood of -0.105 is then selected as a new changepoint τ_p.
2. The changepoint τ_p divides the synthetic time series into $s_p = \{t_1, \ldots, t_{440}\}$ and $s_{p+1} = \{t_{442}, \ldots, t_{542}\}$, the log-likelihood of changepoint is then calculated

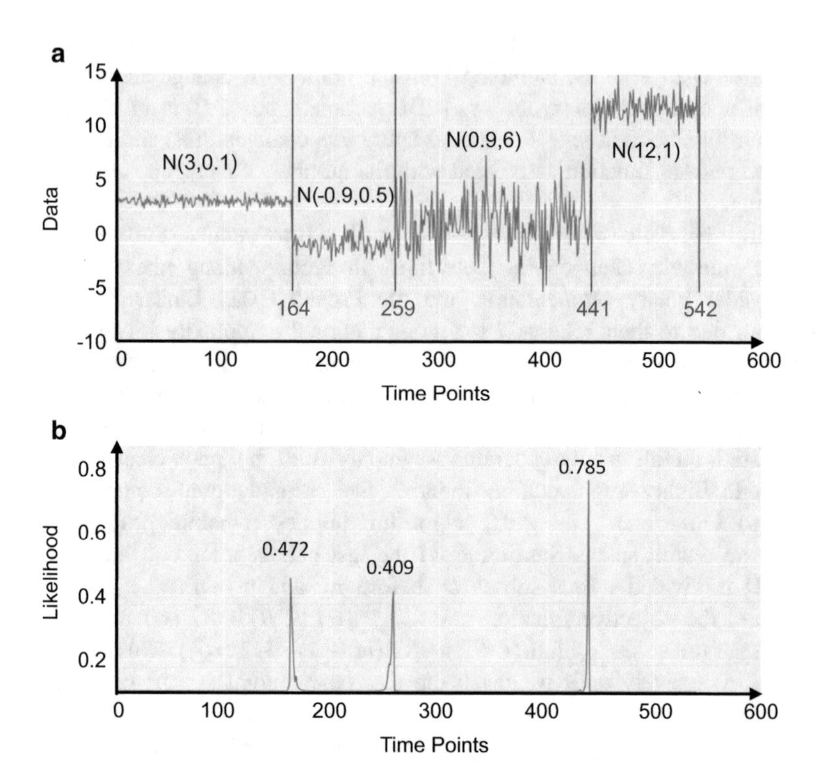

Fig. 3.5 A synthetic example of maximum-likelihood-estimation-based changepoint detection (**a**) Data and estimated locations of changepoints (**b**) The corresponding estimated likelihood associated with changepoints

within s_p and s_{p+1}, respectively. The time point t_{164} in s_p with maximum log-likelihood of -0.326 is thus selected as a new changepoint τ_q.

3. The changepoint τ_q divides s_p into $s_q = \{t_1, \ldots, t_{163}\}$ and $s_{q+1} = \{t_{165}, \ldots, t_{440}\}$. the log-likelihood of changepoint is then calculated within s_q and s_{q+1}, respectively. The time point t_{259} in s_{q+1} with maximum log-likelihood of -0.388 is thus selected as a new changepoint τ_r.
4. Since no changepoint is detected in the newly divided segments, the final identified changepoint list is formed as $CP = \left\{\tau_q = t_{164}, \tau_r = t_{259}, \tau_p = t_{441}\right\}$.

The likelihood of a time point being identified as a changepoint is shown in Fig. 3.5b. We can see that the changepoint likelihood remains low within each segment, but increases significantly between neighboring segments. The time locations of the three peak values of likelihood 0.472, 0.409, 0.785 are therefore identified as changepoints for our synthetic data, which matches the actual data pattern.

3.3.2 Density-Ratio-Estimation-Based Method

Traditional density-estimation-based methods face difficulty in detecting multiple changepoints in long-term multivariate time series. Such difficulty lies in two aspects: (1) high computation cost; (2) limited type of detectable changepoints. The computation cost comes from increasing dimensionality in both temporal and feature domains. In particular, when features are correlated with each other, the computation cost of estimating the joint distribution for the entire feature space can increase exponentially as the feature-space dimensionality increases [7]. Although heuristics algorithms such as PELT can reduce the computation cost, they can only identify limited types of changepoints, e.g., changepoints in mean and variance. Note however that correlation/covariance/frequency changes are common in long-term multivariate time series, but these types of changepoints cannot be effectively captured by traditional density-estimation-based methods [7, 8]. Therefore, a density-ratio-estimation-based approach has been proposed to address this problem [7, 9].

The key idea of the density-ratio estimation is that instead of directly estimating the probability densities of all segments, this method estimates the ratio of probability densities between neighboring segments. This is because the objective of estimating probability densities is to find locations where the difference between the densities of neighboring segments is maximized. Therefore, an alternative solution is to directly estimate the ratio of probability densities between neighboring segments, which can significantly reduce the computation cost [7, 9]. Assuming that s_i and s_{i+1} are two consecutive segments, a dissimilarity measure can be defined as:

$$Divergence(i, i+1) = \int p(s_{i+1}) f(\frac{p(s_i)}{p(s_{i+1})}) ds_i \tag{3.4}$$

where $p(s_i)$ and $p(s_{i+1})$ are the probability density functions corresponding to segments s_i and s_{i+1}. Note that f is a convex function and $f(1) = 0$. Two types of the function f are widely used: (1) Kullback–Leibler (KL) divergence, which has $f(t) = t \log t$; (2) Pearson (PE) divergence, which has $f(t) = \frac{1}{2}(t-1)^2$. The dissimilarity measure for these two types of divergence can thus be written as:

$$KL(i, i+1) = \int p(s_i) \log(\frac{p(s_i)}{p(s_{i+1})}) ds_i \tag{3.5}$$

$$PE(i, i+1) = \frac{1}{2} \int p(s_{i+1}) (\frac{p(s_i)}{p(s_{i+1})} - 1)^2 ds_i \tag{3.6}$$

Next, the density ratio $\frac{p(s_i)}{p(s_{i+1})}$ is modeled by a Gaussian kernel model M:

$$M(s_i, \theta) = \sum_{l=1}^{n} \theta_l K(s_i, s_l) \tag{3.7}$$

$$K(s_i, s_l) = \exp(-\frac{||s_i - s_l||^2}{2\sigma^2}) \tag{3.8}$$

where s_l represents the l-delayed version of s_{i+1} and θ_l is a set of parameters associated with s_l. Note that $K(s_i, s_l)$ is a Gaussian kernel function and σ is the kernel width. Other kernel methods can also be used to boost the performance [10, 11]. After θ_l is learned from training data to minimize the dissimilarity measure, the estimators for the corresponding KL divergence and PE divergence can be calculated as:

$$\widehat{KL} = \frac{1}{n} \sum_{i=1}^{n} \log \widehat{M}(s_i) \tag{3.9}$$

$$\widehat{PE} = -\frac{1}{2n} \sum_{j=1}^{n} \widehat{M}(s_j)^2 + \frac{1}{n} \sum_{i=1}^{n} \widehat{M}(s_i) - \frac{1}{2} \tag{3.10}$$

where s_j is a neighboring segment of s_i. Moreover, since the density-ratio value can be unbounded, a relative density-ratio estimation is used in practice [9]:

$$\widehat{PE} = -\frac{\alpha}{2n} \sum_{i=1}^{n} \widehat{M}(s_i)^2 - \frac{1-\alpha}{2n} \sum_{j=1}^{n} \widehat{M}(s_j)^2 + \frac{1}{n} \sum_{i=1}^{n} \widehat{M}(s_i) - \frac{1}{2} \tag{3.11}$$

3.3.3 *Clustering-Based Method*

Another way to avoid the difficulties associated with density estimation of long-term multivariate time series is to cluster different segments using appropriate divergence metrics so that the difference between clusters can be maximized. Moreover, density-estimation based methods rely on the assumptions that observations are independently and identically distributed and segments obey the Gaussian distribution. Although preprocessing approaches such as the seasonal decomposition can help hold the assumptions true for most cases, there exists situations where these assumptions are not valid. That is why the clustering-based methods are needed to handle more general cases. Hierarchical clustering can be used to iteratively find all locations of potential changepoints [8, 12].

First, a Euclidean-distance-based divergence measure is defined for two segments $s_i = \left\{d_{\tau_{i-1}+1}, \ldots, d_{\tau_{i-1}+n}\right\}$ and $s_j = \left\{d_{\tau_{j-1}+1}, \ldots, d_{\tau_{j-1}+m}\right\}$:

$$\begin{aligned}
\widehat{ED}(s_i, s_j; \alpha) &= 2E|s_i - s_j|^\alpha - E|s_i - s_i'|^\alpha - E|s_j - s_j'|^\alpha \\
E|s_i - s_j|^\alpha &= \frac{1}{mn} \sum_{p=1}^{n} \sum_{q=1}^{m} |d_{\tau_{i-1}+p} - d_{\tau_{j-1}+q}|^\alpha \\
E|s_i - s_i'|^\alpha &= \binom{n}{2}^{-1} \sum_{p=1}^{n-1} \sum_{k=p}^{n} |d_{\tau_{i-1}+p} - d_{\tau_{i-1}+k}|^\alpha \\
E|s_j - s_j'|^\alpha &= \binom{m}{2}^{-1} \sum_{q=1}^{m-1} \sum_{k=q}^{m} |d_{\tau_{j-1}+q} - d_{\tau_{j-1}+k}|^\alpha
\end{aligned} \tag{3.12}$$

where n and m are the lengths of the two segments s_i and s_j, respectively, and α is a predefined constant, $\alpha \in (0, 2]$. Note that the term $E|s_i - s_j|^\alpha$ represents the average Euclidean distance between time points belonging to different segments. The term $E|s_i - s_i'|$ calculates the average Euclidean distance between time points belonging to the segment s_i and the term $E|s_j - s_j'|$ expresses the average Euclidean distance between time points belonging to the segment s_j. The scaled sample divergence measure is then calculated for all neighboring segments [8]:

$$Divergence(i, i+1; \alpha) = \frac{mn}{m+n} \widehat{ED}(s_i, s_{i+1}; \alpha) \tag{3.13}$$

The potential locations of changepoints can be discovered by examining where the $Divergence(i, i+1; \alpha)$ is maximized. However, since the number of changepoints is not given, the number of clusters is also unknown, making traditional clustering algorithms, which require the number of clusters as their prior, infeasible for our problem. In contrast, hierarchical clustering can iteratively find a set of most likely changepoints without knowing the number of clusters in advance.

The divisive and agglomerative methods are two widely used types of hierarchical clustering. The divisive method is a top-down approach where an initial "big" cluster containing all data points is recursively split into two least-similar clusters until there is one cluster for each data point. In contrast, the agglomerative method is a bottom-up approach: each data point starts in its own cluster and pairs of "nearest" clusters are iteratively merged until only a final "big" cluster is left. Figure 3.6 illustrates how the divisive and agglomerative methods can be used to detect changepoints.

1. Divisive method: First, the most likely location of a changepoint in the initial candidate time-series sequence is obtained. This is achieved by examining, using Eq. (3.13), where the divergence measure is maximized. Next, a permutation-based significance test is executed to validate the statistical confidence of this potential changepoint. To achieve this, the original sequence is randomly permuted R times and the corresponding maximum divergence values are then calculated. Note that R is the total number of permutations and the choice of R depends on the distribution of the observations, as well as the number and

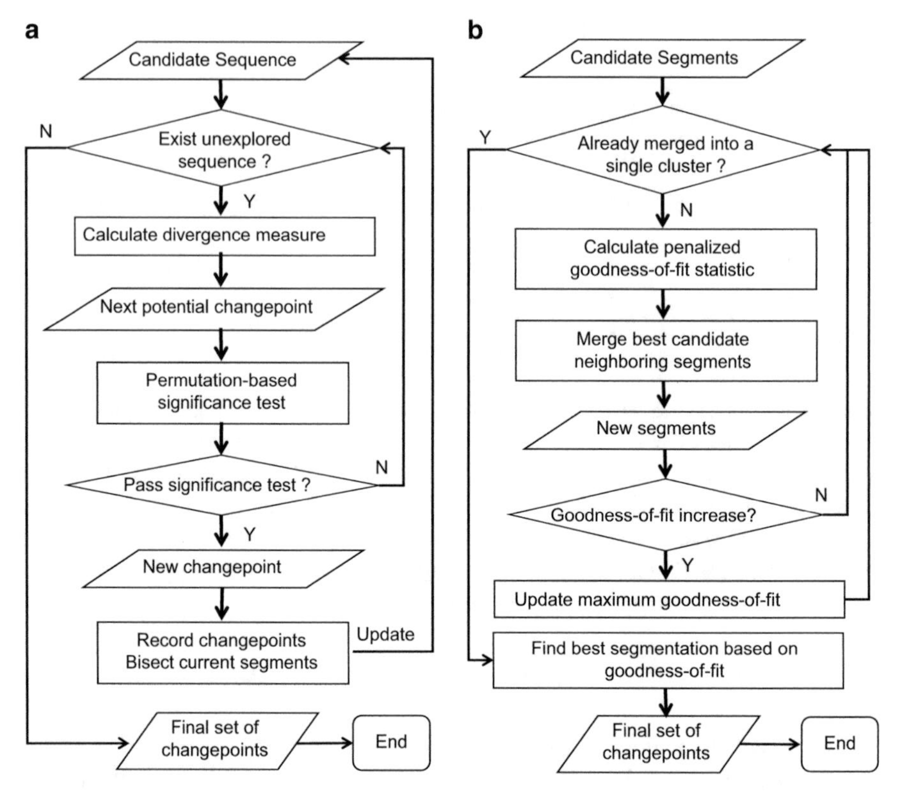

Fig. 3.6 Illustration of clustering-based changepoint detection: (**a**) hierarchical divisive clustering; (**b**) hierarchical agglomerative clustering

size of clusters [8]. The p-value of this significance test is represented by the percentage of permuted sequences that have larger maximum divergence value than the original one. A new changepoint is added only if it has passed this permutation-based significance test. After a new changepoint is identified, it will bisect the current sequence into two new candidate subsequences and the above procedure will be repeated for each of them. The entire hierarchical divisive procedure terminates when no more new changepoints pass the permutation-based significance test. The final set of changepoints is obtained at the end of this procedure.

Agglomerative method: Initially, a set of c neighboring candidate segments $\mathbf{s}$ are chosen from the original time-series sequence. The penalized goodness-of-fit statistic is then calculated to find and merge the best candidate neighboring segments. Specifically, the penalized goodness-of-fit statistic is defined as:

$$GOF_c(\mathbf{s};\alpha) = \sum_{i=1}^{c-1} Divergence(i, i+1;\alpha) + penalty(c) \tag{3.14}$$

where c is the size of the candidate segments. The *penalty* function is associated with the location of changepoints and aims at avoiding over-segmentation. This measure is greedily maximized in each iteration via merging neighboring segments that offer the largest increase or the smallest decrease. The entire hierarchical agglomerative procedure terminates when all candidate segments have been merged into a single cluster. Since the maximum goodness-of-fit statistic has been updated during each iteration, the best segmentation and corresponding set of changepoints will be obtained at the end of this procedure.

3.3.4 Hybrid Method

Even though some of the difficulties involved in detecting changes in long-term multivariate time series can be alleviated by the above methods, different approaches may be sensitive to different types of changepoints. Therefore, a single type of changepoint detection cannot fit the case where a complex combination of different types of changepoints exist. Moreover, experts may have different priorities against different changepoint types, therefore, a filtering component is also needed to enable customizable changepoint analysis.

As shown in Fig. 3.7, the input multivariate time series is fed concurrently to all types of changepoint detectors in the offline component, generating several sets of detected changepoints. A recently proposed Bayesian-based online method (BCPD) is also incorporated here so that we can identify a changepoint as soon

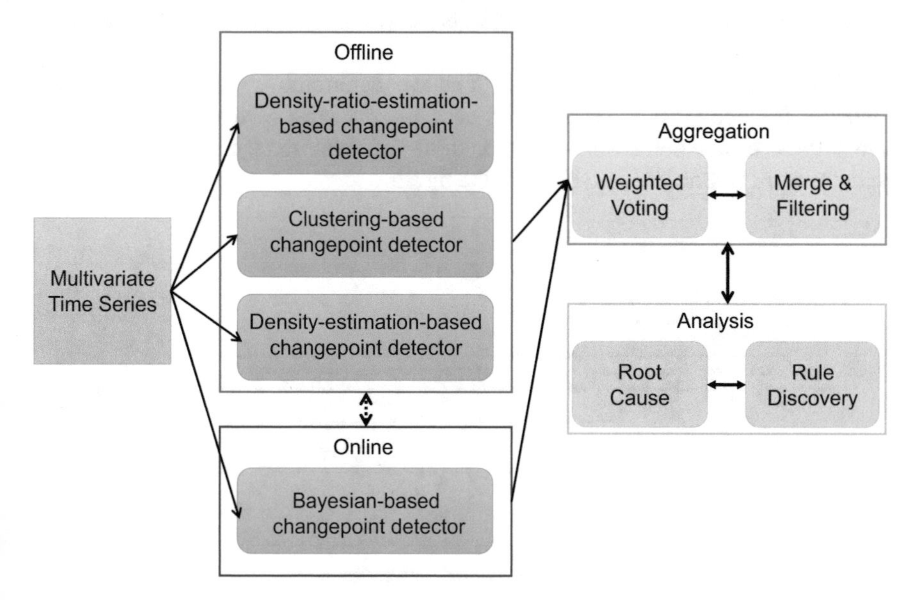

Fig. 3.7 Illustration of hybrid multivariate changepoint detection

as possible [13]. The BCPD method compares the probability distribution of a new sliding window with historical intervals and identifies a changepoint with the highest posterior probability. Changepoints from both offline and online components are fed to the aggregation component. Both weighted voting and user-customized filtering mechanisms are used to select the final set of candidate changepoints. An additional analysis component is also implemented to interpret the identified changepoints, including an analysis of their potential root causes and the discovery of new expert rules.

3.4 Changepoint Window Learning

After all changepoints have been identified, the next step is to identify changepoint windows so that all useful information before and after scenario changes can be incorporated. Assume that a set of changepoints $CP^* = \{\tau_1^*, \ldots, \tau_m^*\}$ has been detected in time-series data D and these changepoints split D into $m + 1$ segments $S^* = \{s_1^*, \ldots, s_{m+1}^*\}$. Now we need to identify a changepoint window w_i for each identified changepoint τ_i. A changepoint window w_i consists of three parts: wb_i, which contains information before τ_i, wa_i, which contains information after τ_i, and the changepoint τ_i itself, as shown in Fig. 3.8. We can see that wb_i is a subset of segment s_i and wa_i is a subset of segment s_{i+1}. Therefore, as long as the lengths of wb_i and wa_i are determined, the data points contained in w_i are also determined. A common choice of length (len) is $len(wb_i) = 0.5 \times len(s_i)$ and $len(wa_i) = 0.5 \times len(s_{i+1})$ so that the final changepoint window w_i not only contains useful information relevant to scenario changes, but is also maintained within a reasonable size [3].

Figure 3.9 shows an example of how we determine changepoint windows. Suppose that the synthetic time series used here is same as that used in Fig. 3.5. After three changepoints are detected, three different changepoint windows are subsequently determined, as shown in the figure.

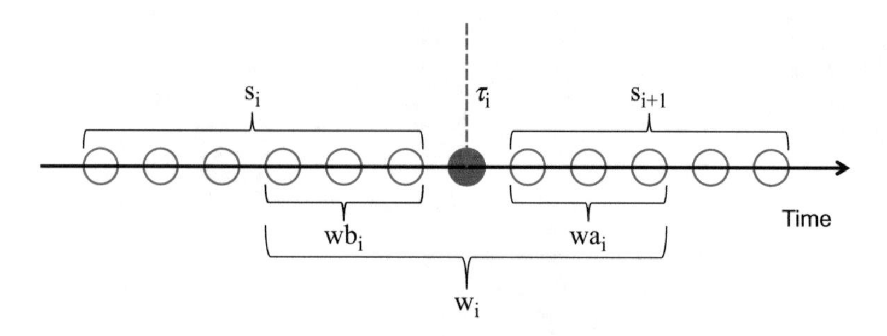

Fig. 3.8 Illustration of a changepoint window in a time series

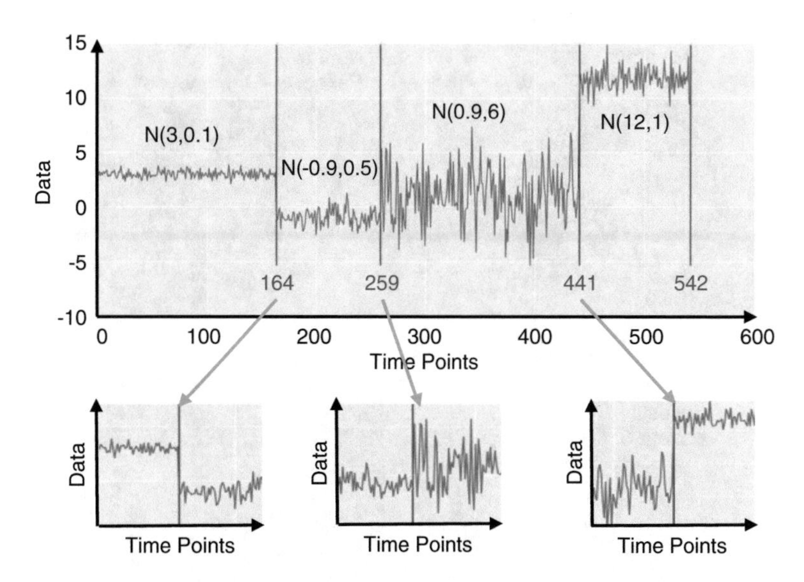

Fig. 3.9 A synthetic example of changepoint window building

After all changepoint windows have been constructed from historical data, we can now apply machine-learning techniques to learn normal/abnormal patterns. Since labels of the CPs are unknown, unsupervised clustering methods are needed to automatically group similar changepoint window patterns. Among different kinds of clustering techniques, density-based spatial clustering of applications with noise (DBSCAN) is promising because it does not require the number of clusters as its prior and it can find arbitrarily shaped clusters [14]. The basic idea of DBSCAN is to form a new cluster if the number of neighboring points is larger than a threshold. More formally, DBSCAN uses two predefined parameters, neighborhood distance ϵ and minimal number of points $minpts$. Using these two parameters, all data points can be classified into three types: (1) *core points*, which have at least $minpts$ number of neighboring points within ϵ distance; (2) *border points*, which are not core points but are neighbors of core points; (3) *noise points*, which are neither core points nor border points. Finally, neighboring core points and their neighboring border points together form a new cluster.

Figure 3.10 shows a simple example of applying DBSCAN to the changepoint window-learning problem. Suppose we have a set of changepoint windows $W = \{w_1, \ldots, w_{14}\}$. Each changepoint window w_i is referred as a point in the figure. If we set $minpts = 3$, two clusters are formed. The cluster in the top-left corner consists of the *core point* set $\{w_3, w_4, w_5, w_{11}\}$ and the *border point* set $\{w_6, w_{10}, w_{12}\}$. The cluster in top-right corner contains the *core point* set $\{w_1, w_7, w_{14}\}$ and the *border point* set $\{w_8, w_{13}\}$.

Note that the DBSCAN algorithm as described in the literature [14] can only identify as anomalies noise points that do not belong to any cluster. However, in

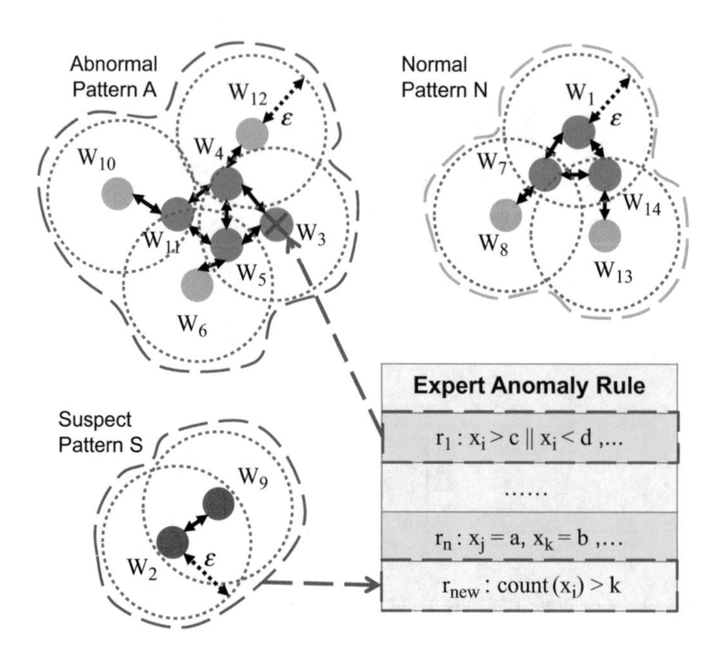

Fig. 3.10 Illustration of DBSCAN-based changepoint window learning

commercial core router systems, during a specific period, some types of normal scenarios may occur rarely while anomalies may occur frequently. In this scenario, a high-density cluster can also represent abnormal patterns and low-density noise points do not necessarily indicate anomalies. Therefore, in this work, we incorporate an updatable *expert anomaly rule table* into the original DBSCAN algorithm so that different types of normal/abnormal window patterns can be learned more accurately. Figure 3.10 shows the improved DBSCAN-based changepoint-window learning method. First, for each cluster identified by DBSCAN, its core point set will be fed to the expert anomaly rule table. If any core point satisfies any of the expert anomaly rules, the entire cluster will be identified as being an abnormal window pattern. For example, the core point w_3 in the top-left cluster satisfies expert anomaly rule r_1. Therefore, the entire cluster is identified as "Abnormal Pattern A". In contrast, none of the core points in the top-right cluster are found in the expert anomaly rule table. Thus, this cluster is labeled as "Normal Pattern N". Moreover, neighboring noise points together are considered as a suspect window pattern. If any matched expert anomaly rules are found, this suspect pattern will be labeled as an abnormal window pattern. Otherwise, a new expert rule will be generated from this suspect pattern and then be inserted into the expert anomaly rule table. For example, the two neighboring noise points in the bottom-left corner form a "Suspect Pattern S". Since no related rules are found in the table, a new rule r_{new} is generated from this pattern and inserted into the table.

3.5 Anomaly Detection

After the library of changepoint window patterns is generated, the next step is to detect anomalies. Each changepoint window pattern p_i in the library can be classified into three general categories: normal, abnormal, and suspect. Assume that the normal pattern category NP has u different window patterns, i.e., $NP = \{np_1, np_2, \ldots, np_u\}$, the abnormal pattern category AP has v window patterns, i.e., $AP = \{ap_1, ap_2, \ldots, ap_v\}$, and the suspect pattern category SP has w window patterns ($SP = \{sp_1, sp_2, \ldots, sp_w\}$). The optimum window size of these patterns is determined during the changepoint-window-learning phase and it is fixed during the anomaly detection phase. When new time-series data $D_{new} = \{d_1, d_2, \ldots, d_l\}$ is extracted from a core router system, it is first fed to the changepoint detection component. If no changepoints are detected, D_{new} is labeled as normal and no further checking is executed. Otherwise, k changepoints given by the set $CP_{new} = \{\tau_1, \tau_2, \ldots, \tau_k\}$ are identified and consequently a set of k changepoint windows $W_{new} = \{w_1, w_2, \ldots, w_k\}$ is determined. For each changepoint window w_i, we have developed two methods to determine its normal/abnormal condition.

1. Distance-based method: The distance $dist(p_j, w_i)$ between w_i and each window pattern p_j in the library is calculated. The window pattern p^* that is closest to w_i can then be obtained by:

$$p^* = \underset{p_j \in NP \cup AP \cup SP}{\arg\min} \; dist(p_j, w_i) \tag{3.15}$$

The neighborhood distance ϵ of p^* is used as a threshold. If the distance between w_i and p^* is smaller than ϵ, w_i is placed in the same category as p^*. Otherwise, w_i is considered to be a new window pattern p_{new} and a cluster checking-updating process is executed. Suppose the neighborhood range of pattern p_j is $r_j = [p_j - \epsilon, p_j + \epsilon]$, then:

- If the neighborhood range r_{new} of the new window pattern p_{new} overlaps with the neighborhood range $R = \left\{r_1, \ldots, r_j\right\}$ of a set of existing patterns $P = \left\{p_1, \ldots, p_j\right\}$, an updated pattern cluster P_{update} is formed $P_{update} = \{p_{new}\} \cup P$. The newly merged cluster is inserted into the abnormal category as long as $\exists p_j$ such that $p_j \in NP$.
- Otherwise, the new window pattern p_{new} is inserted into the suspect pattern category SP.

An example of such a cluster update procedure is shown in Fig. 3.11. Initially, the library contains two clusters: "Abnormal Pattern A" and "Suspect Pattern S". Although the new window w_{15} belongs to neither of these two existing clusters, its neighborhood range r_{15} overlaps with these two clusters. Therefore, a new abnormal pattern cluster A' is formed, where $A' = \{w_{15}\} \cup A \cup S$.

2. Classification-based method: A disadvantage of the distance-based method is that some boundary points belong to different clusters simultaneously, hence it

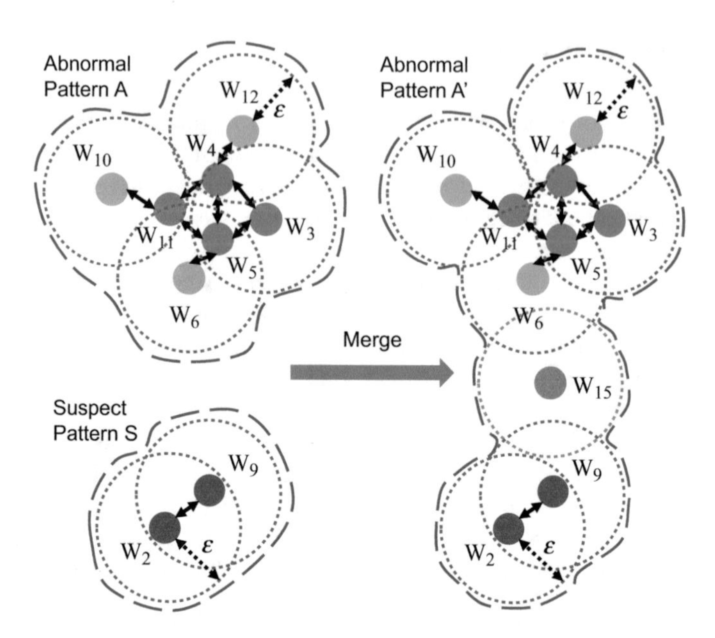

Fig. 3.11 Illustration of cluster merging

is difficult to find a linearly-separable decision boundary. Therefore, we adopt a classification-based method. First, all historical changepoint windows are labeled by their window patterns. The total number of available labels equals the total number of different window patterns $u + v + w$. Then, a machine-learning-based classifier is built using these labeled data. We use support vector machines for this purpose because this approach is able to find an optimal non-linearly-separable decision boundary [15]. Finally, the new window w_i is fed to the trained classifier to predict its label. Besides utilizing a traditional shallow-learning classifier such support vector machines to find an optimal non-linearly-separable decision boundary [15], we also incorporate a deep-learning classifier, namely a recurrent neural network (RNN), to further improve the performance of the proposed method. Unlike traditional feed-forward neural networks, an RNN maintains an internal state (memory) that allows it to preserve dynamic temporal behavior, making it suitable for analyzing time-series data [16, 17]. In addition, the hidden state in RNNs is shared over time and thus it can contain information from an arbitrarily long window [18, 19]. Therefore, the RNN model is promising in learning and classifying a wide range of changepoint window patterns in our classification-based method.

After all changepoint windows in W_{new} are labeled, the overall normal/abnormal condition of the new time series D_{new} can be evaluated. Note that alarms can be reported in either a conservative or an aggressive way. If we choose the conservative method, D_{new} will be reported as anomalous only if at least one of its changepoint windows belongs to the abnormal pattern category. In contrast, if the aggressive method is selected, D_{new} will be reported as anomalous as long as any one of its changepoint windows are labeled as "abnormal" or "suspect".

3.6 Experimental Setup

Twenty commercial core routers with the same platform version and similar hardware configurations were monitored in our experiments. Figure 3.12 shows an example of the "NE40E" core router product that we used. We can see that it consists of a number of different functional units, such as the main processing unit (MPU), line processing unit (LPU), switch fabric unit (SFU), etc. Also, different types of interface and protocols are supported in the "NE40E". A distributed agent-based system has been developed to monitor these twenty core routers in real time. A total of 2892 features were extracted and sampled every 30 min for 30 days of operation, generating a set of twenty multivariate time-series data consisting of 1440 time points. The information regarding features extracted from the different components in core routers is shown in Table 3.1. We can see that different kinds of features are extracted from the various components.

Product	NE40E-X16A
Number of Slots	22 slots: 2 MPUs (1:1 backup), 4 SFUs (3+1 backup), 16 LPUs
Environment	0°C to 45°C
Power Consumption	9040W (480G)
Interface type	100GE/40GE, GE/FE, ...
Software Version	V8
Supported Protocols	IPv4, IPv6, MPLS, ...

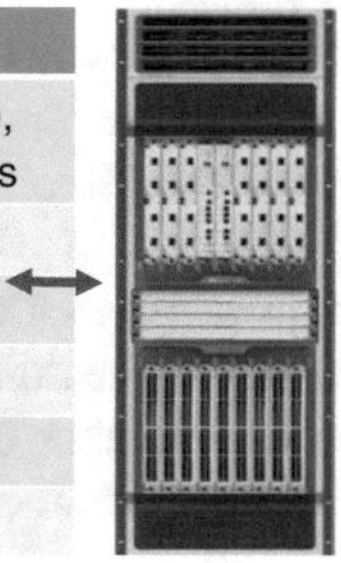

Fig. 3.12 Description of the commercial core router used in our experiments

Table 3.1 Features extracted from core routers

Component	Number of components	Number of features	Representative features
MPU	2 (1 + 1 backup)	112	CPU/Mem usage
			Board temperature
SFU	4 (3 + 1 backup)	196	Route age
			Board uptime
LPU	16	832	Lost packet count
Interface	93	1004	Input/output rate
			Utilization ratio
Others	124	748	Exception/assertion

A fourfold *cross-validation* method [20], which randomly partitions the extracted time series dataset into four groups, is used to evaluate the performance of the proposed changepoint-based anomaly detection method. Each group is regarded as a test case while all the other groups are used for training. We used only fourfold cross-validation because of two reasons. The first is the time cost associated with the training phase. Since grid search is used to determine the optimum changepoint window size, multiple rounds of cross-validation are executed for each parameter setting. Therefore, higher fold cross-validation may increase the time cost significantly. The second reason lies in the limited total number of time-series cases available for our experiments. Therefore, if higher fold cross-validation is used, the number of time series split into the test dataset would decrease, making the validation insufficient. Traditionally, both the accuracy and detection delay are essential metrics to evaluate the effectiveness of various anomaly detection approaches. The accuracy shows whether an approach can identify different types of anomalies while the detection delay indicates whether a method can trigger an alert before severe system failures occur. However, since the current monitoring system alerts experts about anomalies on a per day basis, the time slack is sufficient for the evaluation of offline methods. Therefore, although the total time cost associated with different anomaly detection methods is discussed in Figs. 3.21 and 3.22, which implicitly shows the time-efficiency of different methods, two accuracy-related metrics, namely Success Ratio and Non-False-Alarm Ratio, are used as the main evaluation metrics in this work. The Success Ratio (SR), referred to as a percentage, is the ratio of the number of correctly detected changepoints/anomalies to the total number of changepoints/anomalies in the testing dataset. For example, a SR of 70% means that 7 out of 10 changepoints/anomalies are correctly detected. In addition to SR, the Non-False-Alarm Ratio (NFAR) is also considered as an evaluation metric. It is defined as the ratio of the number of correctly detected changepoints/anomalies to the total number of alarms flagged by the changepoint/anomaly detector. An effective detection method is expected to maximize both SR and NFAR.

3.7 Results of Changepoint Detection

Seven types of changepoints were considered during our experiments. Based on observations from our comprehensive experiments, we conclude that although other types of changepoints also exist, they tend to occur as a combination of these seven basic types.

1. Changepoint in Mean: This type of changepoint occurs when neighboring segments have a significant difference in mean values. Figure 3.13a shows an example of single changepoint in mean value. We can see that the mean value drops approximately from 25 to 10 after it passes the changepoint.
2. Changepoint in Variance: If the variance from the mean is significantly different for two neighboring segments, a changepoint in variance is flagged. An example

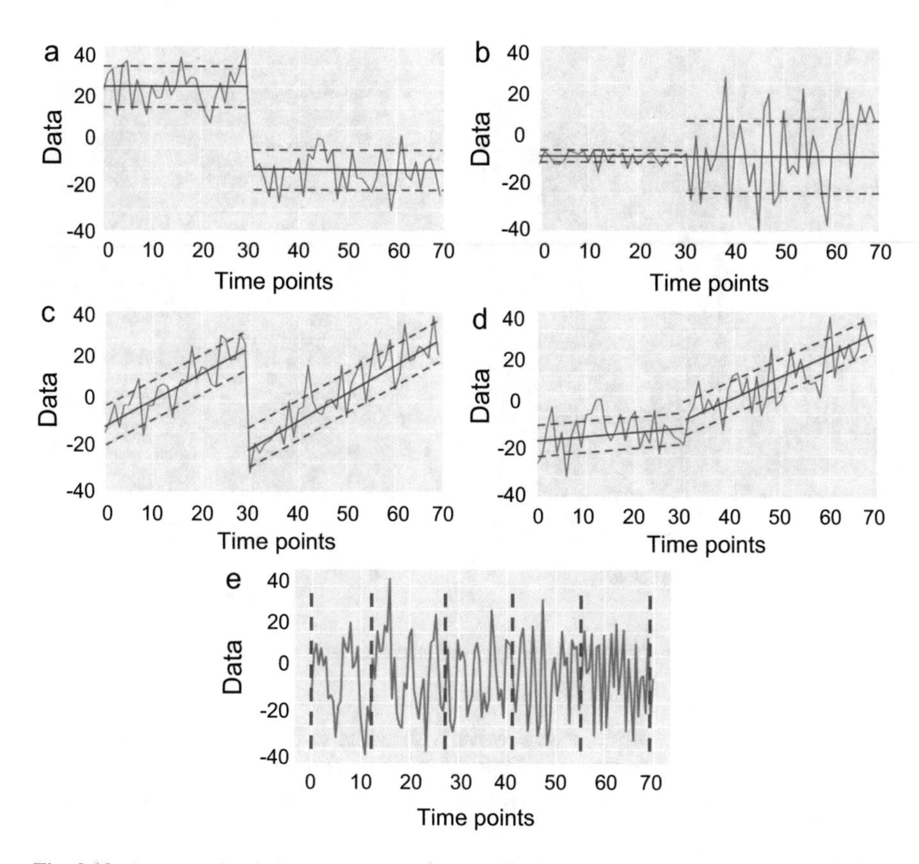

Fig. 3.13 An example of changepoints in (**a**) mean, (**b**) variance, (**c**) intercept, (**d**) trend, and (**e**) frequency

of a single changepoint in variance is shown in Fig. 3.13b. Although the mean value remains constant across the entire temporal domain, the variance of the segment after the changepoint is much larger than the segment before the changepoint.

3. Changepoint in Intercept: When a significant intercept shift is observed between two neighboring segments, a changepoint in intercept is said to occur. Figure 3.13c shows an example of a single changepoint in intercept. Although the gradient remains the same across the temporal domain, the intercept changes from 15 to 0.
4. Changepoint in Trend: A changepoint in trend appears if the trends of two neighboring segments are significantly different from each other. An example of a single changepoint in trend is shown in Fig. 3.13d. It can be observed that the increasing trend after the changepoint is much steeper than the increasing trend before the changepoint.
5. Changepoint in Frequency: When the occurrence rate of repeating patterns for two neighboring segments is significantly different, a changepoint in frequency

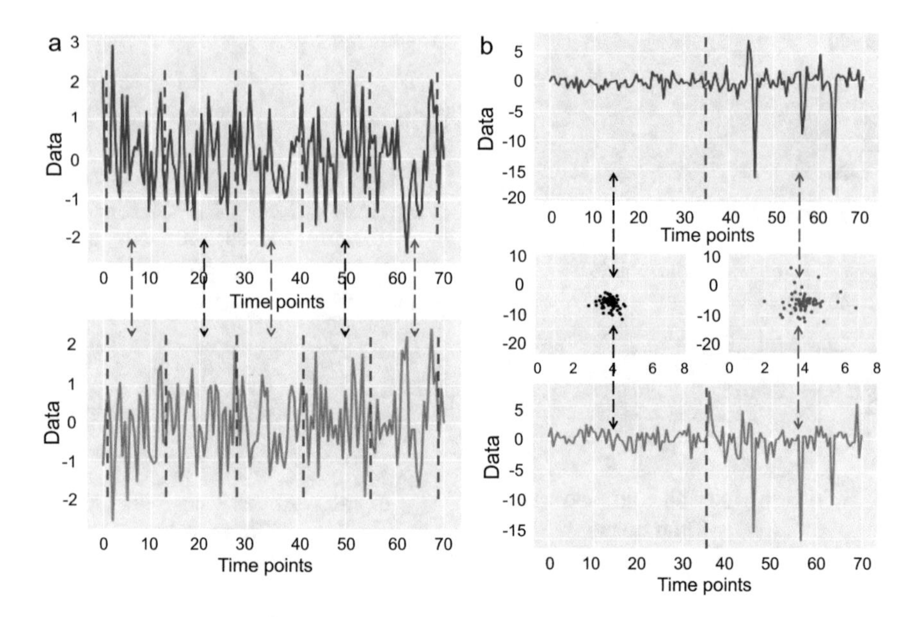

Fig. 3.14 An example of changepoint in (**a**) covariance/correlation, (**b**) tail shape

is observed. For example, a sine wave with white noise is shown in Fig. 3.13e. We can see that the frequency of this sinusoid increases as time proceeds. Thus, segments in the second half exhibit many more repeating patterns than those in the first half.

6. Changepoint in Covariance/Correlation: Features in multivariate time series are usually correlated. Therefore, a changepoint in covariance occurs if the tendency in the linear relationship between the features changes across neighboring segments. An example of a bivariate normal distribution is shown in Fig. 3.14a. It can be seen that the linear relationship between these two features changes from negative correlation to positive correlation in the beginning and goes back to negative correlation in the end.
7. Changepoint in Tail shape: Tail shape can be used to characterize different distributions. A flat distribution with small variance usually has a short tail while a sharp distribution with a high variance often exhibits a long tail. Therefore, a changepoint in the tail is identified if the scatter distribution of neighboring segments is significantly different. For example, the curve in Fig. 3.14b is a mixture of a bivariate normal distribution and a bivariate student-t distribution. We can see that the segment before the changepoint belongs to a normal distribution and it has a relatively dense scatter diagram. In contrast, the segment after the changepoint belongs to a student-t distribution and it has a long tail.

The 3σ rule, where σ refers to the standard deviation, can be used to define the "significant difference" mentioned above: If the mean/variance/intercept/trend of

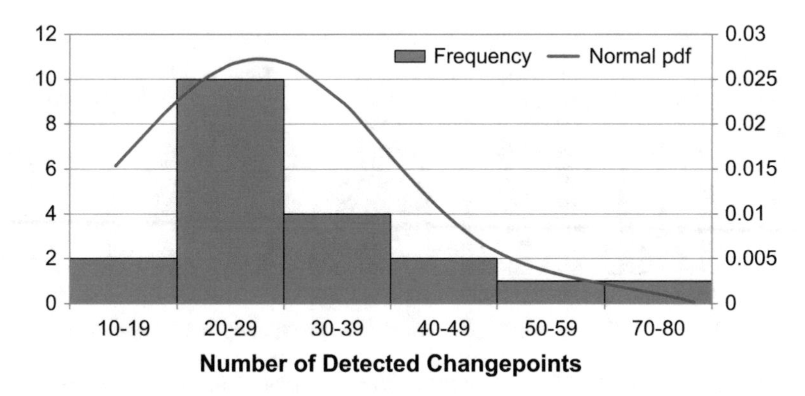

Fig. 3.15 Histogram for the number of detected changepoints

the latter segment lies outside range of that of the former segment, a "significant difference" occurs.

The steps involved in detecting changepoints within a time-series instance are:

1. Apply the changepoint-detection algorithms to the multivariate time series.
2. Record the location τ_i, the type $type_i$, the feature name f_i of each detected changepoint cp_i. Note that the $type_i$ is encoded as $1, 2, 3, 4, 5, 6, 7$ for changepoints in mean/variance/intercept/trend/frequency/covariance/tail.
3. Filter the changepoint set via a predefined minimal window size ml so that two changepoints cp_j and cp_k are merged if $|\tau_j - \tau_k| \leq ml$ and $type_j \neq 6$, $type_k \neq 6$.

Therefore, the final number of detected changepoints is the size of the filtered changepoint set, as shown in Fig. 3.15. We can see that 20–40 changepoints are detected from more than 60% of the time-series instances. Since these 20–40 changepoints are spread throughout the 30-day operation period, per-day scenario changes are indicated, which is in accordance with reality in commercial core routers.

A key objective of our experiment was also to validate the effectiveness of our changepoint detection methods. However, a major problem here is that the ground truth corresponding to the changepoints is unknown. Therefore, heuristic inference is needed to approximate when and where real changepoints occur. Generally, the main scenarios in commercial core routers are determined by running tasks and active configurations. In this case, when the elements contained in the list of running tasks or the active configuration table change significantly, a real changepoint is inferred. If the time distance between an inferred changepoint and a detected changepoint is less than the predefined minimal window size, a match is found. Hence, the success ratio is defined as the fraction of inferred changepoints that can find their matched detected changepoints and the non-false-alarm ratio is the fraction of detected changepoint that can find their matched inferred changepoints.

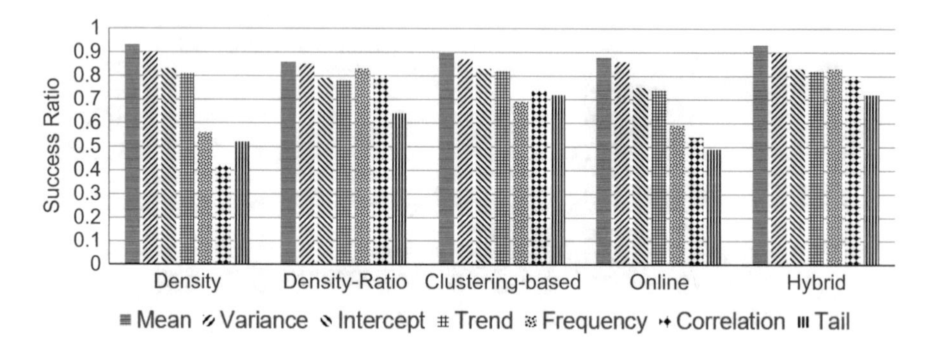

Fig. 3.16 Success ratio of different changepoint detection methods for each changepoint type

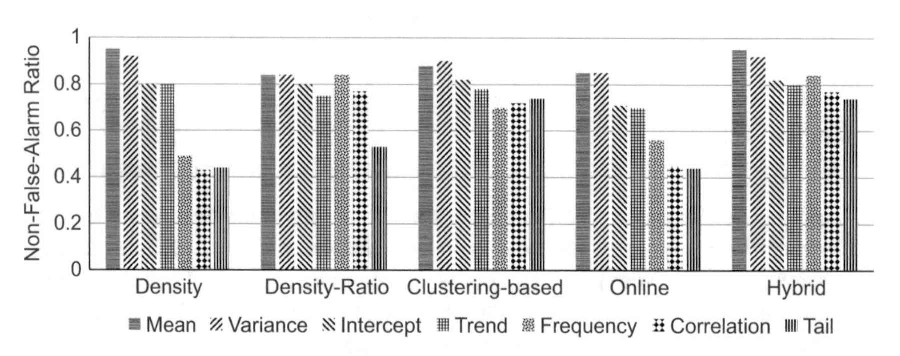

Fig. 3.17 Non-false-alarm ratio of different changepoint detection methods for each changepoint type

Figures 3.16 and 3.17 show the success ratio and non-false-alarm ratio of different changepoint detection methods applied to each changepoint type. The results can be summarized as follows:

1. Density-estimation-based changepoint detection: It achieves the highest success ratio and non-false-alarm ratio for changepoints in mean value and variance value. On the other hand, it performs the worst in detecting changepoints in frequency, covariance, and tail shape. One possible explanation is that density-estimation-based methods such as binary segmentation and PELT are specifically designed for effectively detecting changes in mean and variance. The reason that it can still perform well in detecting changepoints in intercept and trend is because the occurrences of these two types of changepoints often lead to changes in mean or variance as well.
2. Density-ratio-estimation-based changepoint detection: Although it performs a little worse than the density-estimation-based method in identifying changes in mean, variance, intercept and trend, it achieves much higher success ratio and non-false-alarm ratio for changepoints in frequency and covariance. This is because it avoids the high computation cost associated with the estimation of the

joint probability density function of correlated features. Instead, the divergence measure obtained from the estimated density ratio can effectively capture the most likely changepoints in correlations.

3. Clustering-based changepoint detection: It performs well for most types of changepoints. Moreover, it achieves highest success ratio and non-false-alarm ratio for changepoints in tail shape. A possible explanation for this observation is that the characteristics of a statistical tail can be implicitly captured during the hierarchical clustering's iterative procedure.
4. Online Bayesian changepoint detection: As an online algorithm, it can still achieve results that are comparable to the above offline approaches, especially for changepoints in mean and variance. The reason that it does not perform well in detecting covariance and tail changes is due to the insufficient amount of data to fully learn their behaviors.
5. Hybrid changepoint detection: It can be seen that the proposed hybrid method perform best for all types of changepoints. This is because it can always find the most appropriate class of changepoint detector for various changepoints.

Next, the overall success ratio and non-false-alarm ratio of different changepoint detection methods is shown in Fig. 3.18. We can see that for the five changepoint detection methods, i.e., density estimation, density-ratio estimation, clustering-based method, Online Bayesian and the hybrid method, the success ratios are 72.7%, 77.3%, 79.6%, 71.3% and 85.3%, respectively, and the non-false-alarm ratios are 71.6%, 75.7%, 77.1%, 67.1% and 83.4%. The reason that the proposed hybrid method outperform other approaches is that it can overcome the difficulty of matching a single class of changepoint detection to various types of changepoints.

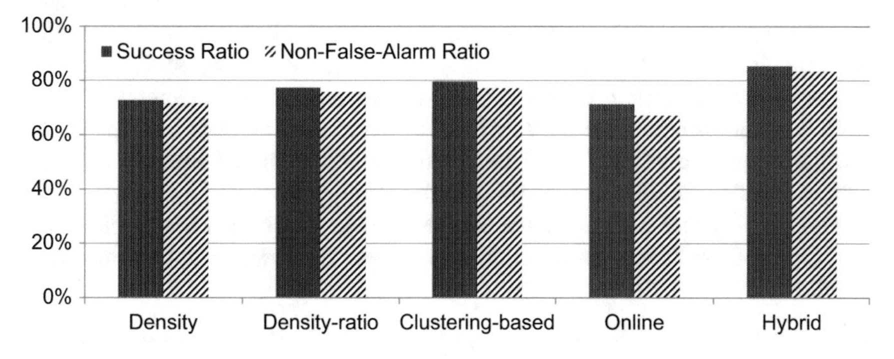

Fig. 3.18 Overall success ratio and non-false-alarm ratio of different changepoint detection methods

3.8 Results of Anomaly Detection

In contrast to the true changepoints that are inferred by heuristic analysis, the true anomalies in our experiments are those historical anomalies that have been identified and labeled by experts in the past. To evaluate the overall effectiveness of the proposed changepoint-based anomaly detection, we have implemented five baseline algorithms. The first baseline ("Rule-based") is a purely rule-based approach. Each rule takes the form "IF $\{\text{feature } f_i = v_i \text{ at time } j\}$, THEN $\{\text{anomaly } a_p = j\}$". For each time point in the testing time-series instance, as long as a matched rule is found in the rule list, a new anomaly is alerted. The second baseline ("KNN") method is a distance-based K-Nearest-Neighbor approach. First, for each testing time-series instance, its distances to all training time-series instances are calculated. Then, it is compared with the training instance with K-nearest distance. A new anomaly alert is generated when the distance difference is larger than a predefined threshold. The third approach ("Win-Clustering") is a window-based clustering method. First, all the training time-series instances are divided into a set of windows $W = \{w_1, \ldots, w_q\}$. Then, a number of clusters $C = \{c_1, \ldots, c_k\}$ are formed from W using the DBSCAN algorithm. Next, each testing time-series instance is divided into a set of windows. A new anomaly alert is generated when a window does not belong to any cluster in C. The fourth baseline ("Raw-RNN") is an RNN-based prediction method. An RNN-based predictive model is first trained using all historical raw time-series data. Next, test time-series instances are divided into two parts. Data points in the first part are fed to the RNN model to generates predicted values. These values are then compared with the actual measured data points in the second part. The accumulated difference between these predicted and the actual observations is defined as the anomaly score for each test time-series instance. The fifth baseline ("Hybrid") is the feature-categorization-based hybrid method from [21]. For the proposed changepoint-based anomaly detection, nearest-neighbor ("CP-1NN") is implemented for the distance-based approach, and support-vector-machine ("CP-SVM") as well as recurrent-neural-network ("CP-RNN") are implemented for the classification-based method. Both the conservative and the aggressive version of the proposed changepoint-based anomaly detection approaches are evaluated during our experiments.

Figures 3.19 and 3.20 show the SR and NFAR values for both baselines and changepoint-based anomaly-detection approaches. The results can be summarized as follows:

1. The distance-based KNN method provides the lowest success ratio and non-false-alarm ratio among all anomaly detection methods. The reason is that the traditional KNN method implicitly assumes that training time-series instances only contain normal scenarios, which is not true in complex core router systems. The data that we have gathered from field operation includes different types of normal and abnormal scenarios that have occurred during the operation of different core routers.

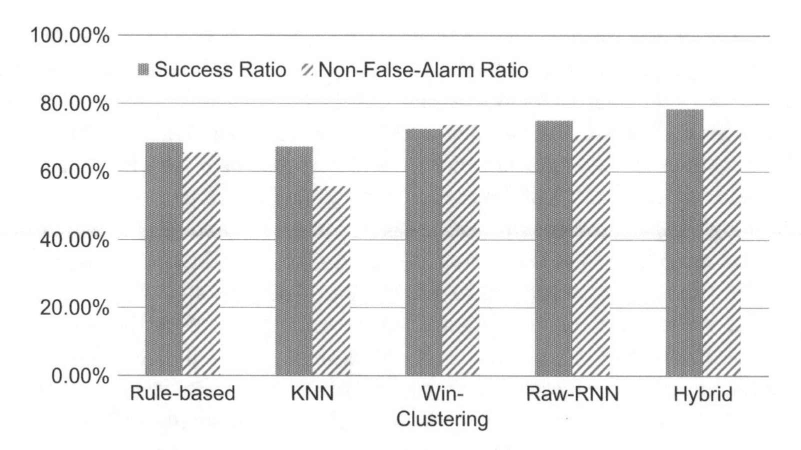

Fig. 3.19 Success ratio and non-false-alarm ratio of baselines

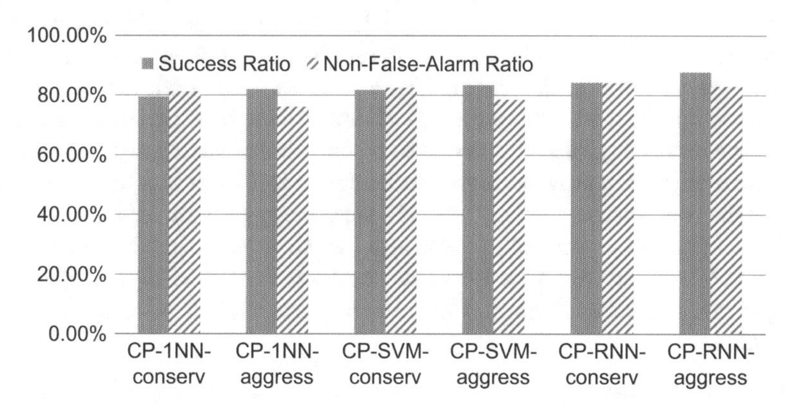

Fig. 3.20 Success ratio and non-false-alarm ratio of changepoint-based anomaly detection methods

2. The purely rule-based approach achieves 68.4% success ratio and 65.5% non-false-alarm ratio. The effectiveness of such a rule-based method depends on whether the rule model covers a sufficient range of "IF features, THEN anomaly" rules. However, there are always new anomalies that do not match any existing rules; therefore, they cannot be detected by the rule-based method.
3. The window-based clustering method performs much better than the KNN and rule-based methods. This is because clustering can capture the most repeatable properties of a subsequence within a time-series instance. However, it still performs worse than the proposed changepoint-based method. One possible reason is that this method assumes that the anomalies are always rare events, which is not the case for complex core router systems. In our experiments, we found that the occurrence frequencies of some anomalies are comparable with that for some normal scenarios.

4. The Raw-RNN method has comparable performance with the window-based clustering approach. This indicates that some irregular temporal behaviors can be captured by the internal memory cell inside the RNN model. However, it achieves lower SR and NFAR than the changepoint-based methods. This is because critical local features for distinguishing normal and abnormal behaviors can be easily lost or distorted in the original long-term multivariate time series.
5. Although the feature-categorization-based hybrid method [21] achieves higher success ratio and non-false-alarm ratio than other baselines, it is outperformed by the proposed changepoint-based method. The reason is that although the hybrid method can choose suitable base anomaly detectors for features with different statistical characteristics, it still assumes that only true anomalies can significantly change the statistical properties of features. However, we found that, for actual data from core routers, a number of statistical changepoints are caused by scenario changes instead of true anomalies.
6. All the three changepoint-based methods: CP-1NN, CP-SVM and CP-RNN achieve higher success ratio and non-false-alarm ratio than the five baselines. The reason for these results are as follows. First, only properties of changepoint windows are considered, reducing false alarms caused by irrelevant or redundant information. Second, different types of abnormal/normal patterns are learned, thereby decreasing false alarms caused by rare normal scenarios. In addition, the aggressive version provides higher success ratio and lower non-false-alarm ratio than the conservative version. This is because the aggressive version also reports suspect patterns as anomalies. Note also that CP-RNN method performs better than CP-1NN and CP-SVM. The reason for this is that CP-RNN can utilize multi-layer non-linear transformation for learning various kinds of complex window patterns.

The time cost associated with all the anomaly detection approaches is shown in Figs. 3.21 and 3.22. We can see that Raw-RNN takes the most time due to the computationally expensive procedure of building and tuning neural nets for long-term multivariate time series. The window-based clustering approach is also time-consuming because during the training phase, it needs to not only determine the

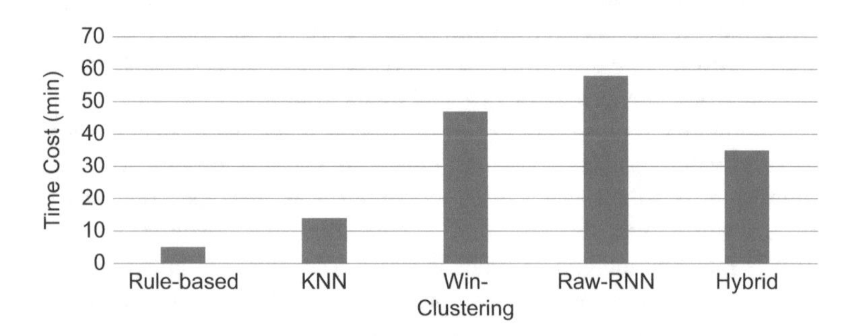

Fig. 3.21 Time cost associated with baselines

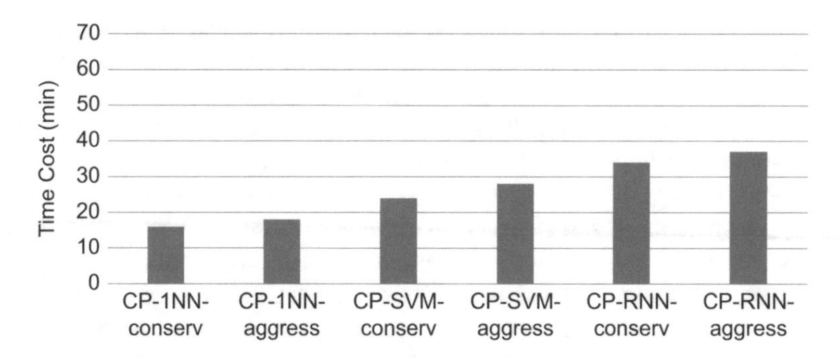

Fig. 3.22 Time cost associated with changepoint-based anomaly detection methods

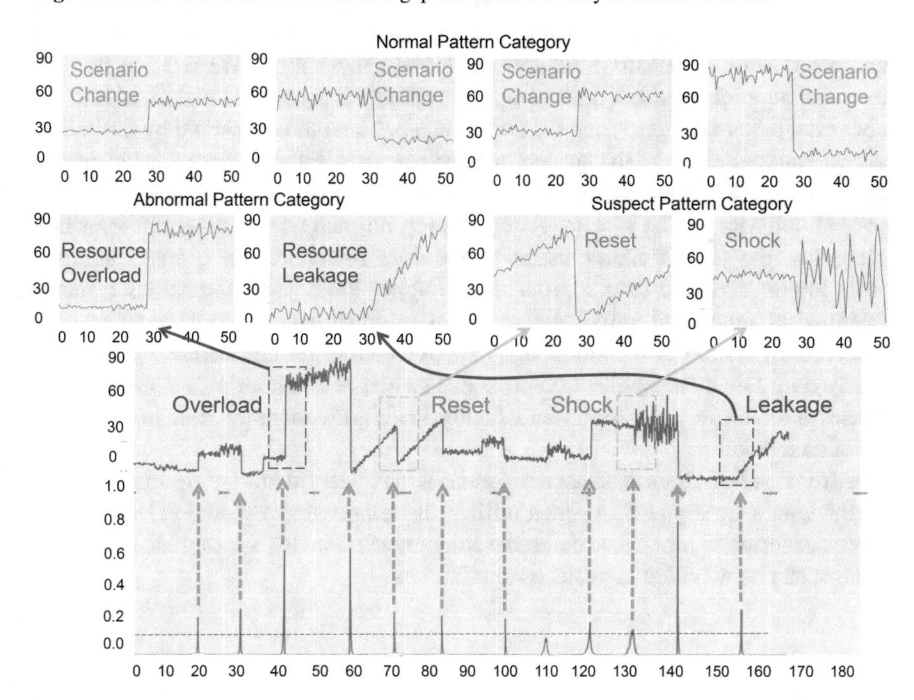

Fig. 3.23 An example of changepoint-based anomaly detection

optimum window size and step size, but also cluster a large number of window candidates. In contrast, the time costs associated with the CP-1NN and CP-SVM are comparable to that for the traditional KNN method and the time cost of CP-RNN is much less than that for Raw-RNN. The reason is that these changepoint-based approaches only need to learn and identify a limited number of changepoint windows.

Figure 3.23 shows an example of our changepoint-based anomaly detection. Assume that four normal patterns of scenario changes, two abnormal patterns

("Resource Overload" and "Resource Leakage"), and two suspect patterns ("Reset" and "Shock"), are learned from the operational data of core routers. We can see that these patterns exhibit different symptoms. For the two abnormal patterns, "Resource Overload" is indicated by the sudden jump in the system's CPU/Memory usage from a low value to the upper bound while "Resource Leakage" occurs if the system's CPU/Memory usage keeps increasing even though no tasks are running. For the two suspect patterns, "Reset" occurs if the trend in the system's CPU/Memory usage is interrupted at some time point but resumes later while "Shock" is indicated by changes in frequency or variance of the system's CPU/Memory usage. Now, a new time series is extracted from a core router system, as shown in the bottom of Fig. 3.23. This data is first fed to our changepoint detection component; a total of 11 changepoints are identified, as marked by the green dotted line in the bottom figure. Next, changepoint windows are determined from these 11 changepoints and compared with the existing window patterns in the library. We can see that the two changepoint windows marked by the red dotted square are classified into the abnormal pattern category, and the two changepoint windows marked by the yellow dotted square belong to the suspect pattern category. Finally, these four identified abnormal/suspect changepoint windows are successfully mapped to an interpretable overall behavior of the system. After running normally for several different task scenarios, the CPU/Memory usage of the core router system changes abruptly from low to high and then approaches the upper limit. Such anomalous resource overload of functional units leads to a system crash and the corresponding reset and reconfiguration of functional units are executed to resume normal operation of the system. Later, the frequent on/off status switch of neighboring network devices causes a shock in the system, and finally leads to a memory leak inside faulty functional units.

After a set of abnormal/suspect scenarios has been identified by the proposed method, an anomaly analyzer then utilizes the temporal and spatial information of these scenarios to root cause defective components. Finally, appropriate preventive actions can be executed to avoid system failures.

3.9 Conclusion

We have described the design of a changepoint-based anomaly detector for a complex core router system. First, different changepoint detection algorithms have been utilized to implement an efficient hybrid changepoint detector. Then, a DBSCAN-based changepoint window learning component has been developed to identify different types of abnormal/normal patterns. Finally, an anomaly detector has been built to alert anomalies in new time-series instances. Data collected from a set of commercial core router systems has been used to evaluate the effectiveness of the proposed methods. Experimental results show that the proposed changepoint-based anomaly detector is significantly more effective and efficient than four baseline methods in detecting anomalies.

References

1. R. Giladi, *Network Processors: Architecture, Programming, and Implementation* (Morgan Kaufmann, Los Altos, 2008)
2. V. Chandola, A. Banerjee, V. Kumar, Anomaly detection: a survey. ACM Comput. Surv. **41**(3), 15 (2009)
3. J. Chen, A.K. Gupta, *Parametric Statistical Change Point Analysis: With Applications to Genetics, Medicine, and Finance* (Springer, Berlin, 2011)
4. J. Aldrich, R.A. Fisher and the making of maximum likelihood 1912–1922. Stat. Sci. **12**, 162–176 (1997)
5. A.J. Scott, M. Knott, A cluster analysis method for grouping means in the analysis of variance. Biometrics **30**(3), 507–512 (1974)
6. R. Killick, P. Fearnhead, I. Eckley, Optimal detection of changepoints with a linear computational cost. J. Am. Stat. Assoc. **107**, 1590–1598 (2012)
7. S. Aminikhanghahi, D.J. Cook, A survey of methods for time series change point detection. Knowl. Inf. Syst. **51**, 339–367 (2017)
8. D.S. Matteson, N.A. James, A nonparametric approach for multiple change point analysis of multivariate data. J. Am. Stat. Assoc. **109**, 334–345 (2014)
9. S. Liu, M. Yamada, N. Collier, M. Sugiyama, Change-point detection in time-series data by relative density-ratio estimation. Neural Netw. **43**, 72–83 (2013)
10. C.S. Teh, C.P. Lim, Monitoring the formation of kernel-based topographic maps in a hybrid SOM-kMER model. IEEE Trans. Neural Netw. **17**, 1336–1341 (2006)
11. G. Bloch et al., Reduced-size kernel models for nonlinear hybrid system identification. IEEE Trans. Neural Netw. **22**, 2398–2405 (2011)
12. J. Cabrieto, F. Tuerlinckx, P. Kuppens, M. Grassmann, E. Ceulemans, Detecting correlation changes in multivariate time series: a comparison of four non-parametric change point detection methods. Behav. Res. Methods **49**, 988–1005 (2017)
13. R. Malladi, G.P. Kalamangalam, B. Aazhang, Online bayesian change point detection algorithms for segmentation of epileptic activity, in *Asilomar Conference on Signals, Systems and Computers* (2013), pp. 1833–1837
14. M. Ester et al., A density-based algorithm for discovering clusters in large spatial databases with noise. **96**, 226–231 (1996)
15. C. Cortes, V. Vapnik, Support-vector networks. Mach. Learn. **20**, 273–297 (1995)
16. M. S. alDosari, Unsupervised anomaly detection in sequences using long short term memory recurrent neural networks. Master's thesis, 2016
17. A. Graves et al., Speech recognition with deep recurrent neural networks, in *International Conference on Acoustics, Speech and Signal Processing* (2013), pp. 6645–6649
18. N. Shone et al., A deep learning approach to network intrusion detection. IEEE Trans. Emerg. Topics Comput. Intell. **2**, 41–50 (2018)
19. W. Wang et al., HAST-IDS: Learning hierarchical spatial-temporal features using deep neural networks to improve intrusion detection. IEEE Access **6**, 1792–1806 (2018)
20. R. Kohavi, A study of cross-validation and bootstrap for accuracy estimation and model selection, in *Proceedings of the International Joint Conference on Artificial Intelligence* (1995), pp. 1137–1143
21. S. Jin et al., Accurate anomaly detection using correlation-based time-series analysis in a core router system, in *Proceedings of IEEE International Test Conference (ITC)* (2016)

Chapter 4
Hierarchical Symbol-Based Health-Status Analysis

To ensure high reliability and rapid error recovery in commercial core router systems, a health-status analyzer is essential to monitor the different features of core routers. However, traditional health analyzers need to store a large amount of historical data in order to identify health status. The storage requirement becomes prohibitively high when we attempt to carry out long-term health-status analysis for a large number of core routers. In this chapter, we describe the design of a symbol-based health status analyzer that first encodes, as a symbol sequence, the long-term complex time series collected from a number of core routers, and then utilizes the symbol sequence to do health analysis. The symbolic aggregation approximation (SAX), 1d-SAX, moving-average-based trend approximation, and non-parametric symbolic approximation representation methods are implemented to encode complex time series in a hierarchical way. Hierarchical agglomerative clustering and sequitur rule discovery are implemented to learn important global and local patterns. Three classification methods including a vector-space-model-based approach are then utilized to identify the health status of core routers. Data collected from a set of commercial core router systems are used to validate the proposed health-status analyzer. The experimental results show that our symbol-based health status analyzer requires much lower storage than traditional methods, but can still maintain comparable diagnosis accuracy.

The remainder of the chapter is organized as follows. Section 4.1 presents the motivation of health status analyzer for core routers. Section 4.2 describes the framework of time-series-based health analyzer. Section 4.3–4.5 present how time series are symbolized, how global and local patterns are learned from symbol sequences, and how different health statuses are identified from pattern library. Experimental results for a commercial core router system are presented in Sect. 4.6. Finally, Sect. 4.7 concludes the chapter.

S. Jin et al., *Anomaly-Detection and Health-Analysis Techniques for Core Router Systems*, https://doi.org/10.1007/978-3-030-33664-6_4

4.1 Motivation

A common way to identify a system's health status is to feed its features to an anomaly detector to see whether any data points are statistical outliers. Besides the traditional distance-based, window-based, and prediction-based anomaly detectors, a feature-categorization-based hybrid method has been presented in Chap. 2 and a changepoint-based anomaly detection has been described in Chap. 3 to detect different types of anomalies within time-varying data streams.

However, an efficient time-series-based anomaly detector is not adequate to obtain a full picture of the health status of monitored core routers. First, an anomaly detector can only provide information about the anomalous points; patterns before or after anomalies are not revealed, which may also be necessary for predicting failures. Second, an anomaly detector can provide little useful information if no anomalies are identified. However, learning different normal patterns are also important because it can reveal how healthy a core router system is and how different task scenarios can affect the system. Therefore, in this work, we address the problem of analyzing the health status of complex core router systems. We use multiple symbolization techniques to discretize and compress long-term time series. We describe several symbol-based clustering and classification methods to analyze the health status of core routers in a more comprehensive way.

4.2 Framework of Time-Series-Based Health Analysis

We propose a time-series-based health-status analysis scheme, as shown in Fig. 4.1. The key idea is that instead of directly analyzing health status from a large volume of raw time series data, we first transform and segment the high-dimensional time series into a meaningful low-dimensional representation. The transformed time series is then fed to our pattern-learning component and health-analysis component for further fine-grained analysis.

1. Pattern-learning component: This component consists of two parts: clustering and rule discovery. The objective of clustering is to group time series with similar shapes. Therefore, a number of different global patterns are learned after clustering. In contrast, the objective of rule discovery is to find repeating local subsequences within long-term time series. Therefore, a number of local patterns are learned after rule discovery.
2. Health-analysis component: This component consists of two parts: classification and prediction. The classification part is used to identify categorical health-status level, and the prediction part is utilized to predict some specific numerical health-status metrics.
3. Expert rule table: This component is maintained and updated by an expert team. The expert rules can be used to either label patterns learned from the pattern-learning component or validate the health status identified by the health-analysis component.

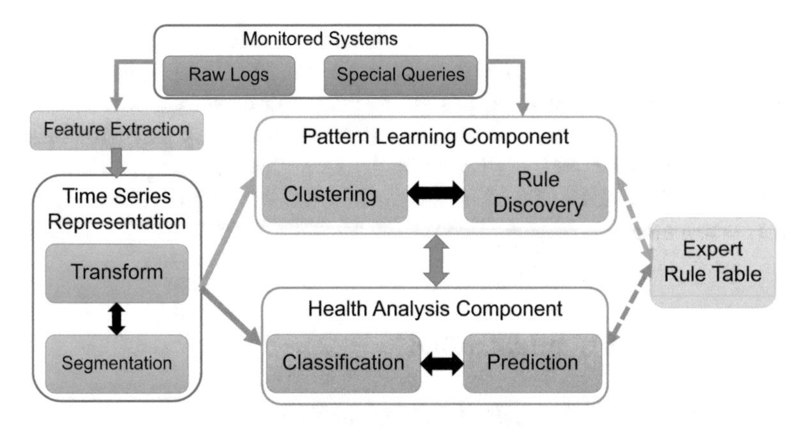

Fig. 4.1 The proposed health-status analyzer in core router systems

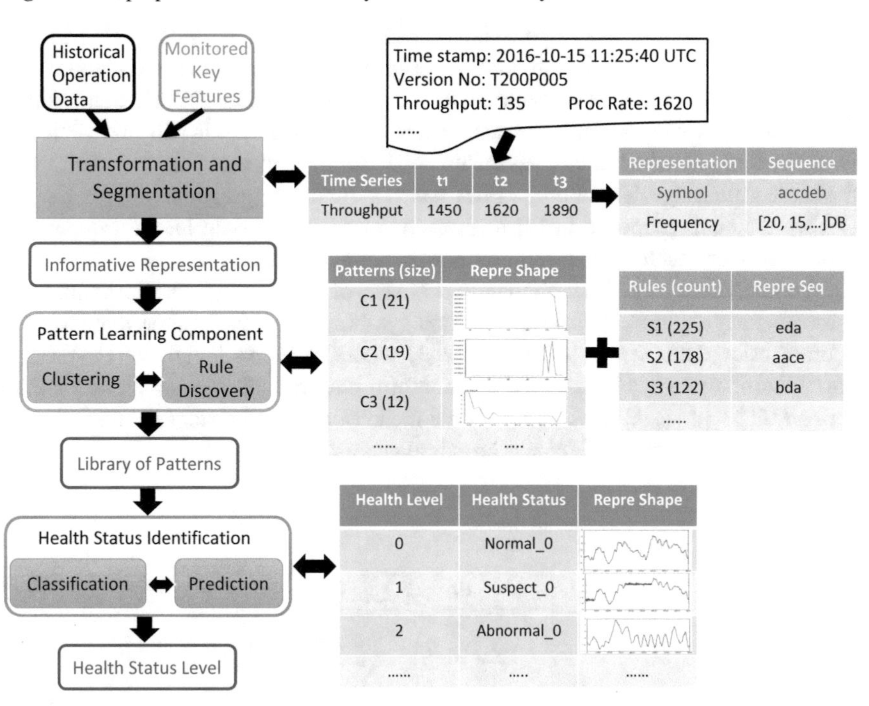

Fig. 4.2 The computation flow of the symbol-based health-status analyzer

Figure 4.2 shows how the proposed health-status analyzer works. We can see that in the beginning, clean and aligned time series for the feature "throughput" are extracted and processed from raw logs. Such high-dimensional time series are then transformed and segmented into a meaningful low-dimensional representation using different techniques such as symbol representation and frequency-domain transform. After the informative representation has been obtained, it is then fed to our pattern-learning component. The pattern library containing all kinds of critical health-status indicators are learned in this component. For example, in Fig. 4.2, three different global representative shapes are formed after clustering while three local representative rule sequences are obtained after rule discovery. After the library of patterns have been learned, the health status identification component is used to classify different levels of health status so that the full picture of the monitored systems is revealed.

4.3 Time Series Symbolization

4.3.1 Symbolic Aggregate Approximation

The piecewise aggregation approximation (PAA) method is widely used to discretize time-series data [1]. It divides the original time series into a set of equal-size segments, and then represents each segment by the mean value of all elements within that segment. However, since the mean value of each segment is still real-valued, it is difficult to tell whether two segments with "similar" mean value truly belong to the same data range. Therefore, a PAA-based symbolic aggregation approximation (SAX) has been proposed in [1] to discretize time-series data into a sequence of symbols, where each symbol represents a predefined data region.

The procedure of SAX symbolization consists of three main steps: normalization, piecewise aggregation approximation, and symbol mapping. Assume that we have a time-series data $D = \{d_1, d_2, \ldots, d_n\}$, where n is the length of D. First, a normalization process is executed to transform the original time series into a time series $DN = \{dn_1, dn_2, \ldots, dn_n\}$ with a mean of zero and a standard deviation of one. Each element dn_i in DN can be calculated as:

$$
\begin{aligned}
dn_i &= \frac{d_i - \mu}{\sigma} \\
&= \frac{d_i - \frac{1}{n}\sum_{i=1}^{n} d_i}{\sqrt{\frac{1}{n}\sum_{i=1}^{n} d_i^2 - \left(\frac{1}{n}\sum_{i=1}^{n} d_i\right)^2}}
\end{aligned} \tag{4.1}
$$

Next, a piecewise aggregation approximation is used to convert the n-length normalized time series DN to a w-length discretized time series $DP = \{dp_1, dp_2, \ldots, dp_w\}$. Each element dp_i in DP can be calculated as:

$$
dp_i = \frac{1}{s} \sum_{j=s(i-1)+1}^{min(si,n)} dn_j \tag{4.2}
$$

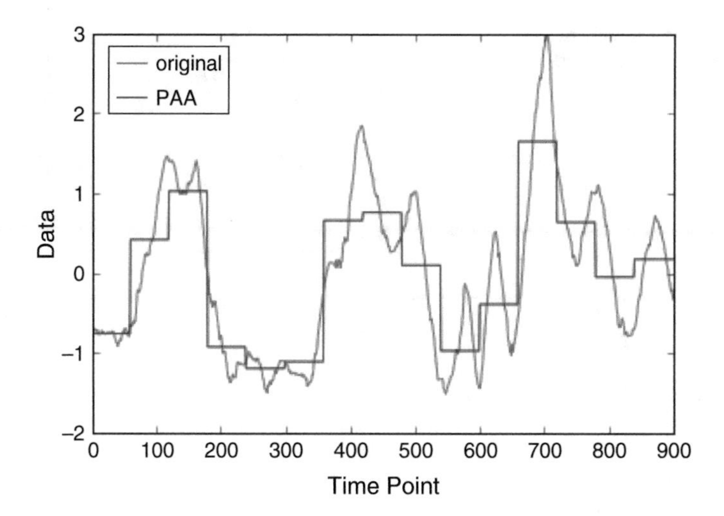

Fig. 4.3 Illustration of piecewise aggregation approximation (PAA)

where $s = \lfloor \frac{n}{w} \rfloor$. Figure 4.3 shows an example of PAA discretization. We can see that PAA partitions the original time series into 15 segments where each segment contains 60 time points. Since each segment is represented by the mean value of all elements within that segment, a 15-length discretized sequence is formed, i.e., $DP = (-0.763, 0.418, 1.024, \ldots, 0.172)$. Therefore, a 1:60 compression ratio is achieved and the global shape of the original time series is preserved.

The final step is to map the w-length discretized time series DP into a w-length symbol sequence $DS = (ds_1, ds_2, \ldots, ds_w)$. The key idea here is to map "close" discretized data points into the same data region, and each data region is represented by a unique symbol. Since the original time series D has been normalized in the first step, its data region can be divided into α equal-sized areas based on the statistical probability property of the Gaussian distribution. A list of $\alpha + 1$ "breakpoints" $B = \{b_0, b_1, b_2, \ldots, b_{\alpha-1}, b_\alpha\}$ can be precomputed to help partition the $N(0, 1)$ Gaussian curve into α equal-sized areas. Given that the breakpoint list is sorted in ascending order, $b_0 = -\infty$, $b_\alpha = \infty$, the value of each breakpoint b_i can then be computed via iteratively solving the following equation:

$$\begin{aligned} \int_{b_i}^{b_{i+1}} N(0, 1) &= \int_{b_i}^{b_{i+1}} \frac{1}{\sqrt{2\pi}} e^{-\frac{x^2}{2}} \\ &= \frac{1}{\alpha} \end{aligned} \tag{4.3}$$

We can see that both the length of B and the value of b_i depend on the choice of the parameter α. Table 4.1 shows an example of breakpoints computed using different values of α ranging from 3 to 7.

Table 4.1 An example of breakpoints calculated with different values of α

$b_i \backslash \alpha$	3	4	5	6	7
b_1	−0.43	−0.67	−0.84	−0.97	−1.07
b_2	0.43	0	−0.25	−0.43	−0.57
b_3		0.67	0.25	0	−0.18
b_4			0.84	0.43	0.18
b_5				0.97	0.57
b_6					1.07

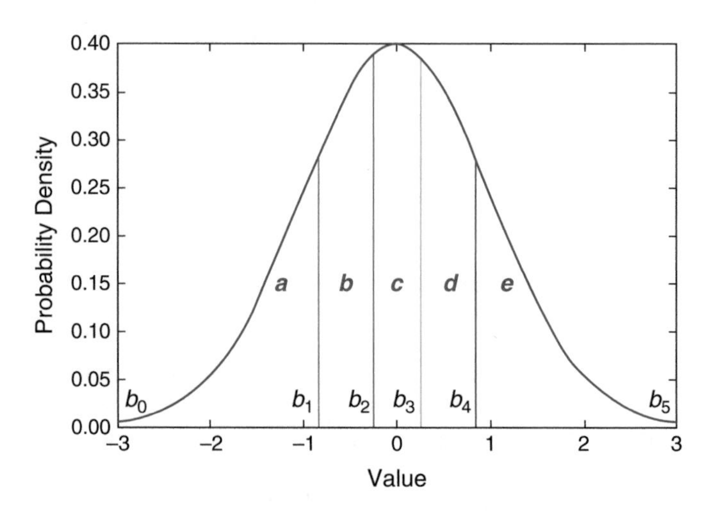

Fig. 4.4 Illustration of symbol mapping in SAX

Once the values of all breakpoints are obtained, partitioning the normal distribution curve becomes a trivial task. For example, Fig. 4.4 shows how a normal distribution can be divided into five equal-sized area using a set of six breakpoints $B = \{-\infty, -0.84, -0.25, 0.25, 0.84, \infty\}$. The resulting equal-sized areas are then encoded by unique characters in a sorted English alphabet $L = \{l_1 = a, l_2 = b, l_3 = c, \ldots, l_{25} = y, l_{26} = z\}$. For example, in Fig. 4.4, the five equal-sized areas are represented by the characters "a", "b", "c", "d", "e", respectively. Now, for any normalized PAA data point dp_i, the symbol ds_i assigned to it can be calculated as:

$$ds_i = l_j, \qquad \text{if} \quad b_{j-1} \leq dp_i \leq b_j \tag{4.4}$$

If the above symbolization procedure with parameter $\alpha = 5$ is applied to the sample time series shown in Fig. 4.3, a 15-length symbol sequence is formed, i.e., $DS = (bdeaaaddcabedcc)$

4.3.2 1d Symbolic Aggregate Approximation

An important advantage of the SAX representation is that it can transform a long-term complex time series into a much shorter and concise symbol sequence without losing the global shape information of the original data. However, important local information such as shapes and trends within a segment are lost in traditional SAX because each symbol in the SAX representation contains only the mean value information of its corresponding segment. A simple example is shown in Fig. 4.5. We can see that the segments in Fig. 4.5a, b are encoded into the same SAX symbol "a" due to their similar average value. However, their trends are significantly different: the segment in Fig. 4.5a is increasing while the segment in Fig. 4.5b is decreasing. Therefore, an improved symbolic representation for time series, called 1d-SAX, has been recently proposed to combine both the average and the trend information of each segment [2].

The key idea of 1d-SAX is to quantize the linear regression of each segment into a symbol and integrate it into the original SAX representation. Linear regression is widely used to model linear relationships between variables and the 1d-SAX approach utilizes it to estimate the trend line of each segment across the temporal domain. Assume that we have normalized time-series data $DN = \{dn_1, dn_2, \ldots, dn_n\}$ with a mean of zero and a standard deviation of one. Similar to the traditional SAX procedure, such an n-length normalized time series DN is divided into a w-length ($w \ll n$) discretized sequence $DP = \{dp_1, dp_2, \ldots, dp_w\}$. Each segment dp_i in DP is represented by a pair (k_i, μ_i), where k_i is the estimated trend value and μ_i is the estimated mean value. To obtain the value of k_i and μ_i, appropriate linear regression is needed to minimize the distance between estimated values and observed values. To achieve this, a linear function $f(t) = kt + b$ is first defined, where k is the slope value and b represents the intercept value. A least square estimation is then executed to find the optimal k_i and b_i for segment dp_i:

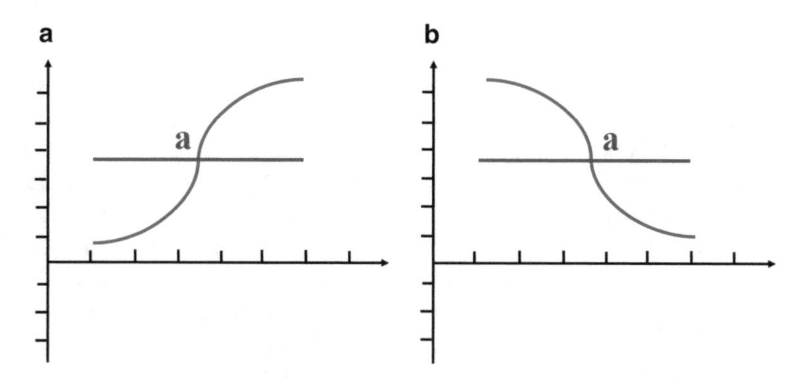

Fig. 4.5 Illustration of segments with the same SAX symbol but different trends

$$
\begin{aligned}
k_i &= \frac{\sum_{j=s(i-1)+1}^{min(si,n)} (t_j - \overline{t}) dn_j}{\sum_{j=s(i-1)+1}^{min(si,n)} (t_j - \overline{t})^2} \\
b_i &= \frac{1}{s} \sum_{j=s(i-1)+1}^{min(si,n)} dn_j - k_i \times \overline{t}
\end{aligned}
\tag{4.5}
$$

where $s = \left\lfloor \frac{n}{w} \right\rfloor$ and $\overline{t} = \frac{1}{s} \sum_{j=s(i-1)+1}^{min(si,n)} t_j$. The estimated mean value μ_i is then calculated as:

$$
\mu_i = \frac{1}{2} \times k_i \times (t_{s(i-1)+1} + t_{min(si,n)}) + b_i \tag{4.6}
$$

After we have obtained the estimated trend k_i and mean μ_i for each segment dp_i, the next step is to quantize these two values and transformed them into a SAX symbol. Since the estimated trend and mean are calculated from linear regression of a normalized time-series segments, these two metrics also obey the Gaussian distribution. Specifically, $k_i \sim N(0, \sigma_s^2)$ and $\mu_i \sim N(0, 1)$, where σ_s^2 is a function of the length of each segment s, and can be approximated as $\sigma_s^2 = 0.03/s$. Therefore, the data region of k_i and μ_i can be divided into β and α equal-sized areas using a list of precomputed "breakpoints". The list of $\alpha + 1$ breakpoints for μ_i can be obtained via Eq. (4.3). Since σ_s^2 is determined by the length of each segment s and β is a predefined parameter that indicates the size of alphabet, similar equation can be used to iteratively compute all breakpoints $\{b_j, j = 0, 1, \ldots, \beta\}$ for k_i, as shown in Eq. (4.7).

$$
\begin{aligned}
\int_{b_j}^{b_{j+1}} N(0, \sigma_s^2) &= \int_{b_j}^{b_{j+1}} \frac{1}{\sqrt{2\pi}\sigma_s} e^{-\frac{x^2}{2\sigma_s^2}} \\
&= \frac{1}{\beta}
\end{aligned}
\tag{4.7}
$$

Once the values of all breakpoints for μ_i and k_i are obtained, the resulting equal-sized areas are encoded by unique characters in a sorted English alphabet: $L = \{l_1 = a(A), l_2 = b(B), \ldots, l_{26} = z(Z)\}$. To distinguish between symbols for μ_i and k_i, lowercase letters are assigned to μ_i and uppercase letters are used to encode k_i. The final 1d-SAX symbol is then obtained by combining the letters for μ_i and k_i. Figure 4.6 shows an example of such symbol mapping in 1d-SAX. We can see that six breakpoints divide the distribution of μ_i and k_i into five equal-sized data regions. The equal-sized data regions of μ_i and k_i are encoded by the characters "a", "b", "c", "d", "e" and "A", "B", "C", "D", "E" respectively. A 5×5 symbol mapping table is then formed to represent different combinations of symbols for μ_i and k_i. Now, for any segment dp_i, after determining its data region of μ_i and k_i, its 1d-SAX symbol can be obtained via looking up such a symbol mapping table.

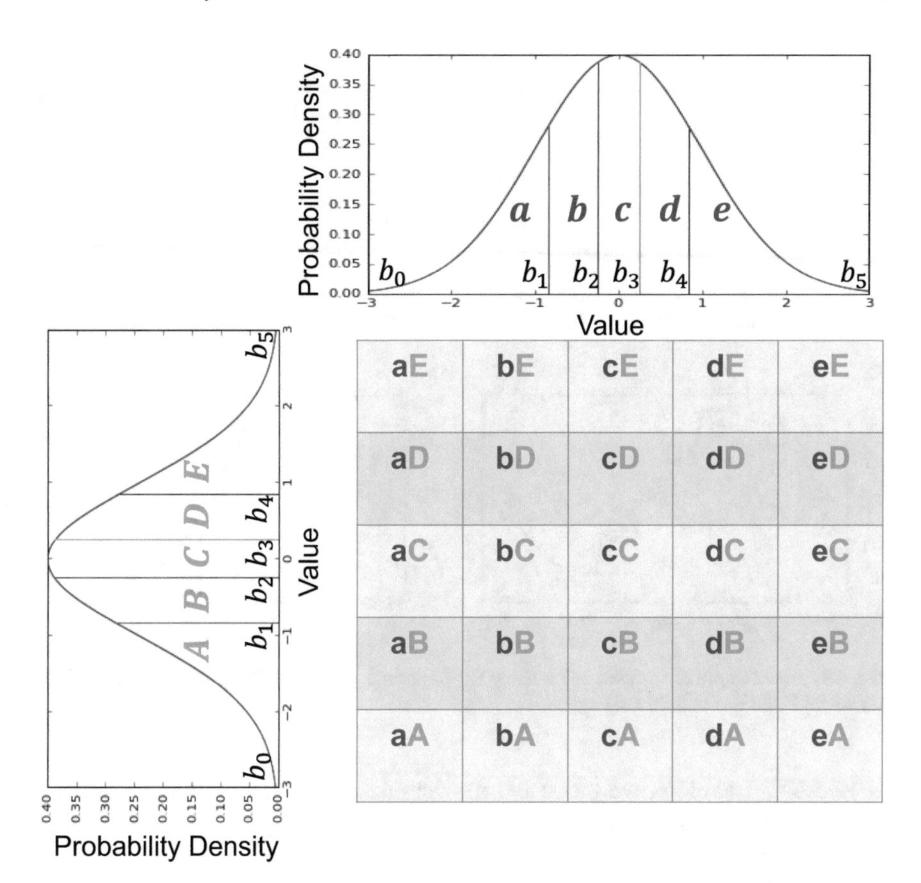

Fig. 4.6 Illustration of symbol mapping in 1d-SAX

An example of symbolization using PAA, SAX, and 1d-SAX is shown in Fig. 4.7. We can see that for the original time series shown in Fig. 4.7a, its PAA and SAX representation are $(0.0, 0.9, 1.0, 1.4, 0.7, 0, -0.9, -1.3, -1.4, -0.4)$ and $(d, g, g, h, g, d, b, a, a, c)$, respectively, as shown in Fig. 4.7b, c. Both PAA and SAX preserve only the mean-related information. In contrast, as shown in Fig. 4.7d, additional trend-related information within each segment is indicated in the 1d-SAX method and the obtained symbol sequence is $(dF, gF, gF, hB, gA, dA, bA, aA, aF, cF)$.

4.3.3 Moving-Average-Based Trend Approximation

Although the 1d-SAX approach can incorporate local linear trend information within each segment, some complex trends such as zigzag are still difficult to capture

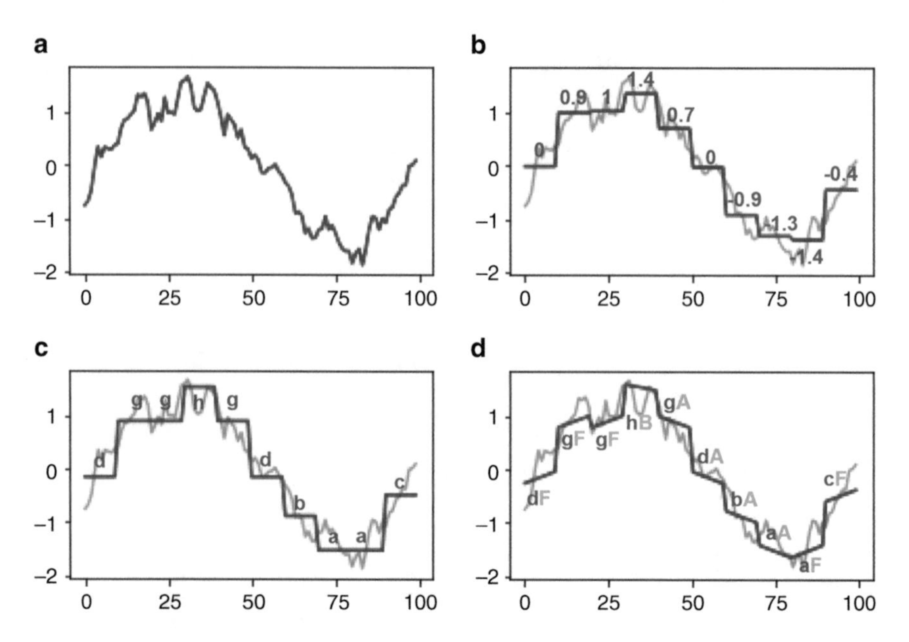

Fig. 4.7 An example of symbolization using (**a**) Raw time series. (**b**) PAA. (**c**) SAX, 8 symbols. (**d**) 1d-SAX, 48 symbols (8 × 6)

in 1d-SAX. Therefore, we propose a moving-average-based trend approximation method to extract, transform, and preserve critical local trend information within a long-term time series.

The moving average (MA) is a commonly used technique to smooth the original real-valued time series. The basic idea of MA is to represent the original time series by mean values calculated from a set of overlapped data sub-windows. More formally, assume that we have a time-series data $D = \{d_1, d_2, \ldots, d_n\}$, where n is the length of D. The window size ω and step size s serve as the parameters for our moving-average method. A set of sliding windows is then formed $W = \{w_1, w_2, \ldots, w_p\}$, where $w_i = \{d_{1+(i-1)s}, \ldots, d_{1+(i-1)s+\omega}\}$. The mean value μ_i for each sliding window w_i can be calculated as:

$$\mu_i = \frac{1}{\omega} \sum_{j=1+(i-1)s}^{1+(i-1)s+\omega} d_j \tag{4.8}$$

Therefore, the original time series D is now smoothed as $U = \{\mu_1, \mu_2, \ldots, \mu_p\}$. The parameter window size ω determines how abrupt changes are handled in the original time series. The larger the window size ω is, the smoother the post-processing time series will become. For example, for the original time series (blue line) shown in Fig. 4.8, three parameter settings are used: $\omega = 2, 4, 6$. We can

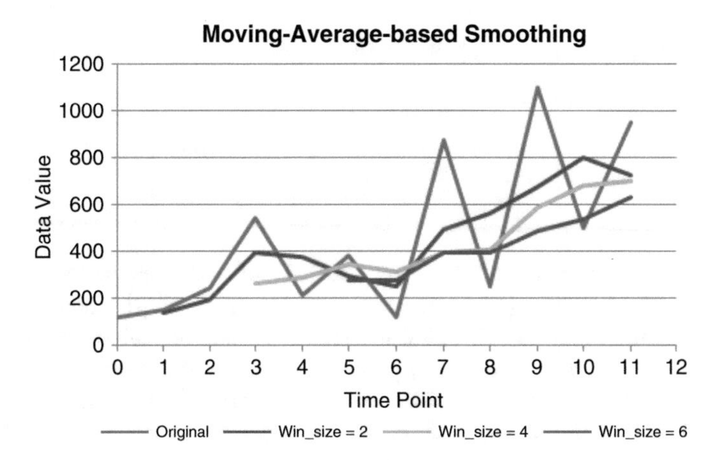

Fig. 4.8 Illustration of moving-average-based smoothing with different window size

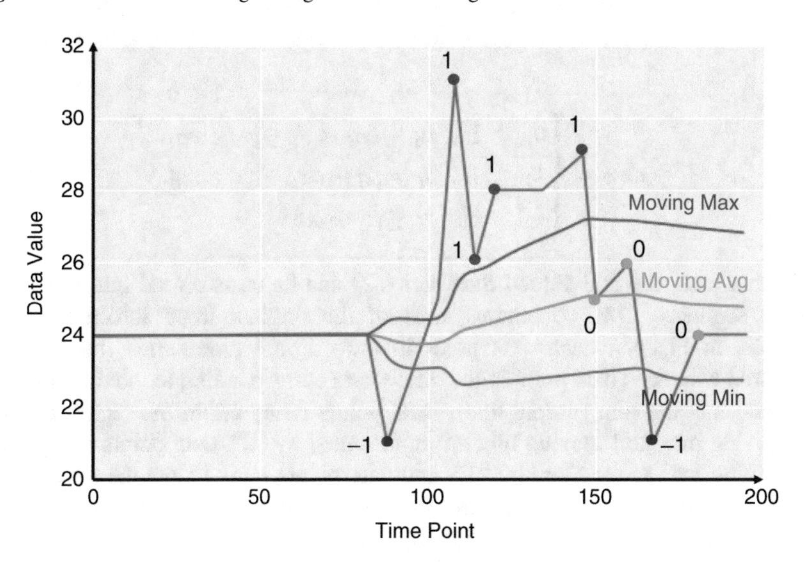

Fig. 4.9 Illustration of moving-average-based trend approximation

see that the post-processing time series with $\omega = 6$ (purple line) is more smooth compared with the other two post-processed time series.

After moving-average-based smoothing, the next step is to identify local trend information of a time series with symbol sequences. Here, we want to represent a time series with a sequence of digits via the 3σ rule, as shown in Fig. 4.9. The 3σ rule, where σ refers to the standard deviation, can be used to define a band around the mean in a normal distribution with a width of weighted standard deviations [3]. To achieve this, the upper bound and lower bound of the smoothed time series are first calculated via the 3σ rule. Assume that the variance set for the sliding window set W is $V = \left\{\sigma_1^2, \sigma_2^2, \ldots, \sigma_p^2\right\}$, where the variance σ_i^2 for each sliding window w_i is calculated as:

$$\sigma_i^2 = \frac{1}{\omega} \sum_{j=1+(i-1)s}^{1+(i-1)s+\omega} d_j^2 - \left(\frac{1}{\omega} \sum_{j=1+(i-1)s}^{1+(i-1)s+\omega} \right)^2 \tag{4.9}$$

Therefore, the lower bound (moving min) L of the time series D can be calculated as $L = \left\{\mu_1 - c\sigma_1, \mu_2 - c\sigma_2, \ldots, \mu_p - c\sigma_p\right\}$, and the upper bound (moving max) H of the time series D can be calculated as $H = \left\{\mu_1 + c\sigma_1, \mu_2 + c\sigma_2, \ldots, \mu_p + c\sigma_p\right\}$. Note that the weight parameter c represents the number of standard deviations used to determine the lower and upper bounds. The higher the value of c is, the wider is the interval between the lower and upper bounds. The next step in this procedure is to utilize a trinary representation to transform the time series D into a digit symbol sequence $DM = \left(dm_1, dm_2, \ldots, dm_p\right)$, where each digit symbol dm_i can be defined as:

$$dm_i = \begin{cases} 0, & \text{if} \quad \mu_i - c\sigma_i \le d_i \le \mu_i + c\sigma_i. \\ 1, & \text{if} \quad d_i \ge \mu_i + c\sigma_i. \\ -1, & \text{if} \quad d_i \le \mu_i - c\sigma_i \end{cases} \tag{4.10}$$

In this case, the real-valued time series D can be transformed into the simple trinary sequence DM containing most of the original trend information. For example, in Fig. 4.9, each data point in the original time series (blue line) is compared with each data point in the time series corresponding to moving max (red line) and moving min (purple line). Data points lying within the region between the moving max and moving min are represented by "0", data points lying above the moving max are marked by "1", and data points lying below the moving min are represented by "−1". Therefore, the final transformed digit sequence can be calculated as $DM = (-1, 1, 1, 1, 1, 0, 0, -1, 0)$.

Besides the transformed trinary digit sequence, several summary metrics can also be calculated to better represent the original time series. For example, an increasing ratio can be used as a magnitude-independent metric to indicate the fraction of data points identified as "increasing" (flag = 1), and it can be directly computed from our trinary sequence. In contrast, the absolute difference, which represents the magnitude difference across the trend domain, is a magnitude-dependent metric that can only be obtained with the help of the original time series. For instance, since the digit sequence obtained in Fig. 4.10 is $DM = (0, 0, 1, 1, 1, 1, 1, 1)$, the increasing ratio can be directly computed as $6/8 = 75\%$. In contrast, for the absolute difference, the magnitudes of the peak and valley need to be first obtained from the original time series, which are 16 and 7 respectively, and then the metric can be calculated as $16 - 7 = 9$.

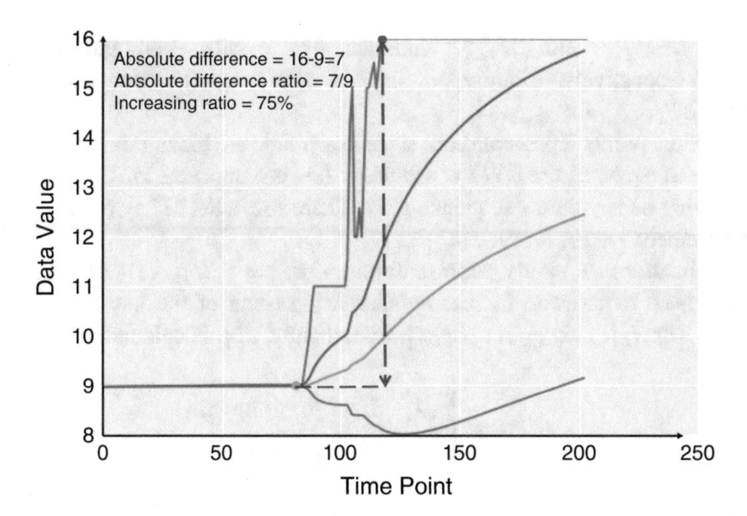

Fig. 4.10 Illustration of metrics used in moving-average-based trend identification

4.3.4 Non-parametric Shape Approximation

However, the above three time series symbolization approaches are parametric techniques, and their effectiveness depends on whether suitable parameter settings such as the number and length of segments can be found, which is difficult and computationally expensive for complex long-term time series. Therefore, a recently proposed non-parametric symbolic approximation representation (NSAR) based on discrete wavelet transform (DWT) can be applied here to encode characteristics of complex time series with little prior information [4].

The key idea of NSAR is that instead of directly symbolizing the raw time series, it first extracts key points of wavelet approximate coefficients that retain critical shape information, and then symbolizes these transformed sequence. The detailed procedure of NSAR can be summarized as follows:

1. Extract wavelet approximate coefficients: a wavelet transformation is first applied to the raw time series, extracting both low-frequency coefficients LF_1 and high-frequency coefficients HF_1. Since high-frequency components usually consist of noises or disturbing factors while low-frequency components retain multi-level critical characteristics, LF_1 is further divided into a new low-frequency part LF_2 and high-frequency part HF_2. This process is repeated until we reach the $\log_2 n$ decomposition level, where n is the length of the raw time series. The coefficients LF_i and HF_i at decomposition level i can be approximated as:

$$
\begin{aligned}
LF_i &= \frac{LF_{i-1}^{(even)} + LF_{i-1}^{(odd)}}{\sqrt{2}} \\
HF_i &= \frac{LF_{i-1}^{(even)} - LF_{i-1}^{(odd)}}{\sqrt{2}}
\end{aligned}
\quad (4.11)
$$

where $LF_{i-1}^{(even)}$ and $LF_{i-1}^{(odd)}$ indicate LF_{i-1} with even and odd sample index, respectively. Finally, a list of DWT coefficients are extracted: $\{LF_1, LF_2, \ldots, LF_{\log_2 n-1}\}$.

2. Extract key points representation: since key points are local maximum/minimum point of a sequence, the DWT coefficients LF_i is compressed in this step via only preserving its key points sequence KP_i. Therefore, a list of key points sequences are obtained: $\{KP_1, \ldots, KP_{\log_2 n-1}\}$.
3. Symbolization: A binary representation $DB_i = \{db_{i1}, \ldots, db_{ij}, \ldots, db_{im}\}$ is first used to indicate increasing/decreasing trend of the key point sequence $KP_i = \{kp_{i1}, \ldots, kp_{ij}, \ldots, kp_{im}\}$. Specifically, db_{ij} is calculated as:

$$db_{ij} = \begin{cases} +1, & \text{if} \quad kp_{ij} \geq kp_{i(j-1)} \\ -1, & \text{if} \quad kp_{ij} < kp_{i(j-1)} \end{cases} \tag{4.12}$$

Next, since -1 and 1 appear alternately in DB_i, the first symbol db_{i1} and the length m can be used to compress DB_i into a new integer $DI_i = m \times db_{i1}$. In this way, the final NSAR symbol sequence is formed: $DI = \{DI_1, \ldots, DI_{\log_2 n-1}\}$.

4.4 Symbol-Based Pattern Learning

After the original high-dimensional time-series data has been transformed to a low-dimensional symbol sequence, clustering and rule discovery algorithms can be applied to learn different patterns from these symbol sequences.

4.4.1 Hierarchical Clustering

Flat clustering such as k-means is efficient for partitioning the original dataset into a specific number of disjoint clusters. However, it treats all clusters equally, and thus does not reveal useful structure information within and among clusters. This drawback makes flat clustering unsuitable for analyzing long-term complex time-series data. A pair of long-term time-series instances sharing similar global shape can still present notable differences in local patterns while two time-series instances having significantly different trends can also contain several identical local patterns. In contrast, hierarchical clustering can not only divide the dataset into a set of clusters, but also output a dendrogram, which is a tree diagram to illustrate the arrangement of the clusters [5].

The divisive and agglomerative methods are two widely-used types of hierarchical clustering. The *divisive* method is a top-down approach where an initial "big" cluster containing all data points is recursively split into two least-similar clusters until there is one cluster for each data point. In contrast, the *agglomerative*

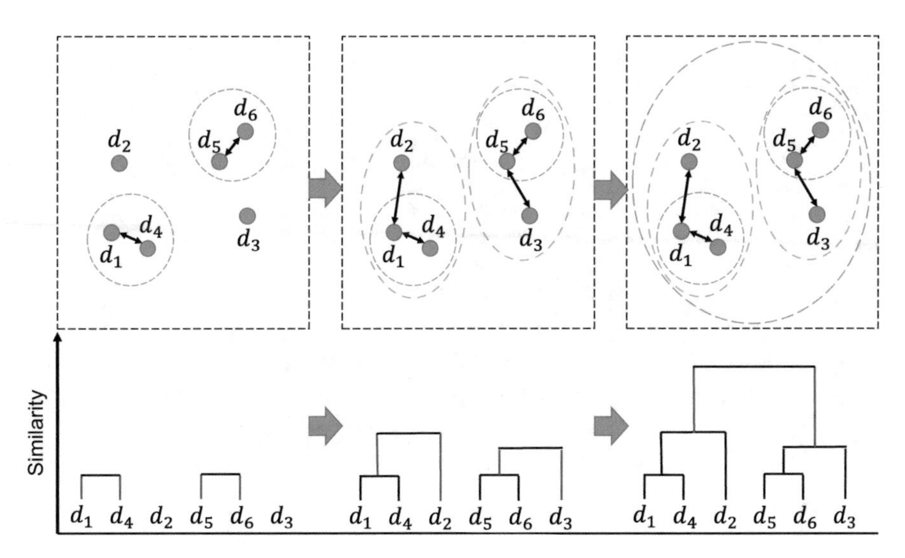

Fig. 4.11 Illustration of hierarchical agglomerative clustering

method is a bottom-up approach: each data point starts in its own cluster and pairs of "nearest" clusters are iteratively merged until only a final "big" cluster is left. Figure 4.11 illustrates how hierarchical agglomerative clustering works. Initially, since the dataset D contains six data points $D = \{d_1, d_2, d_3, d_4, d_5, d_6\}$, six single-element clusters are formed. Next, two pairs of nearest single-element clusters are merged to form two larger clusters: (d_1, d_4), and (d_5, d_6) respectively. The two single-element clusters d_2 and d_3 are then combined with their nearest neighboring clusters, forming two three-element clusters: (d_1, d_2, d_4) and (d_3, d_5, d_6). Finally, these two three-element clusters are merged into a "big" cluster containing all the six data points. Moreover, Fig. 4.11 also shows how a dendrogram is iteratively built via hierarchical clustering. We can see that when smaller clusters are merged into larger clusters, a new level of cluster hierarchy is formed in the dendrogram. Therefore, the final dendrogram generated in Fig. 4.11 indicates a three-level cluster for the sample dataset D.

4.4.2 Rule Discovery

Hierarchical clustering is efficient in building a multi-level structure of clusters, which enables us to analyze and label global patterns from different granularities. However, patterns generated from clustering are sometimes overly complex, which masks some simple but truly meaningful local patterns. Therefore, the rule-discovery component is necessary to learn patterns or rules that are hidden or not obvious in time-series data.

Action	Symbol Sequence	Digram	Rule Utility
Start Scanning	S -> bdeaaaddcabedcc	bd, de, ea, aa, aa, ad, dd, dc, ca, ab, be, ed, dc,cc	
A new rule R1 is made	S -> bdeaaaddcabedcc R1 -> dc	bd, de, ea, aa, aa, ad, dd, dc, ca, ab, be, ed, dc, cc	R1 : 0
Substitue "dc" with the new rule "R1"	S -> bdeaaad"R1"abedcc R1 -> dc	bd, de, ea, aa, aa, ad, dR1, R1a, ab, be, ed, dc, cc	R1 :1
Substitue "dc" with the rule "R1"	S -> bdeaaad"R1"abe"R1"c R1 -> dc	bd, de, ea, aa, aa, ad, dR1, R1a, ab, be, eR1, R1c	R1 :2

Fig. 4.12 Illustration of sequitur rule discovery

We utilize the sequitur algorithm [6] to discover hidden patterns because it can learn a hierarchical structure of rules from any discretized symbol sequence in linear time. The basic idea of sequitur is to recursively substitute repeating subsequence in the input symbol sequence with new rules and thus reconstruct the symbol sequence with a hierarchical-rule-based representation. Two constraints, namely digram uniqueness and rule utility, are used in the sequitur algorithm to improve its efficiency. A *digram* is a subsequence of two adjacent elements from a symbol sequence, and digram uniqueness ensures that no diagram occurs more than once in the symbol sequence via the replacement of both occurrences of the diagrams with a new rule. To satisfy the constraint of rule utility, after replacing all diagrams with new rules, another scanning of the new symbol sequence is executed to ensure that any rule that only occurs once is removed and replaced with symbols that create it. Figure 4.12 presents a simple example that shows how the sequitur algorithm works when we apply it to the symbol sequence generated in Fig. 4.3, which is $DS = (bdeaaaddcabedcc)$. We can see that a new rule $R1 \rightarrow dc$ is extracted when the digram "dc" is found to occur twice in DS. All occurrences of "dc" are then substituted with the new rule $R1$. A final check ensures that the constraints of digram uniqueness and rule utility are both satisfied. Therefore, the rule $R1 \rightarrow dc$ is discovered from the symbol sequence and the new hierarchical-rule-based symbol sequence is $\left(bdeaaad``R1''abe``R1''c\right)$.

4.5 Symbol-Based Classification and Prediction

After informative patterns and rules have been learned from historical time-series data, efficient classification and prediction techniques are needed to identify or predict the health status of new time-series data with these learned patterns.

4.5.1 Distance-Based Method

Distance-based techniques are widely used to classify time series, where the overall "distance" between a pair of time-series instances is used to measure the similarities between them [7]. Therefore, time-series instances are classified into the same category if the overall "distance" between them, $Dist(i, j)$, is less than a threshold t. The key challenge now is on how to choose a proper distance measure to distinguish different time series. The Euclidean distance is commonly used for real-valued time series; it is defined as the square root of the sum of the squared differences of each pair of corresponding data points. For time-series instances symbolized by the SAX method, a specific distance function MINDIST is defined to mimic the Euclidean distance defined on the original real-valued time series [1]. Given two w-length symbol sequences $DS1$ and $DS2$ generated from n-length time-series instances, the minimum distance, or MINDIST, is defined as:

$$\text{MINDIST}(DS1, DS2) = \sqrt{\frac{n}{w}}\sqrt{\sum_{i=1}^{w}(dist(ds1_i, ds2_i))^2} \tag{4.13}$$

where $dist(ds1_i, ds2_i)$ calculates the distance between two symbols via a lookup table generated by the precomputed breakpoint set $B = \{b_0, b_1, b_2, \ldots, b_{\alpha-1}, b_\alpha\}$. If the two symbols are neighborhoods in the alphabet $L = \{l_1 = a, l_2 = b, \ldots, l_{26} = z\}$, their distance $dist(ds1_i, ds2_i) = 0$. Otherwise, let us assume that $ds1_i = l_j$ and $ds2_i = l_k$. The distance is then calculated as $dist(ds1_i, ds2_i) = b_{\max(j,k)-1} - b_{\min(j,k)}$, where the values of $b_{\max(j,k)-1}$ and $b_{\min(j,k)}$ are looked up from the precomputed breakpoint set B. For example, if $DS1 = (bdeaaaddcabedcc)$ and $DS2 = (ccccdeebaababee)$, and MINDIST is recalculated as:

$$\text{MINDIST}(DS1, DS2) = \sqrt{\frac{900}{15}}\sqrt{\sum_{i=1}^{15}(dist(ds1_i, ds2_i))^2}$$

$$= \sqrt{60} \times \sqrt{0.59^2 \times 5 + 0.5^2 \times 2 + 1.68^2 \times 2 + 1.09^2}$$

$$= 23.33$$

After the distance between symbol sequences is successfully represented, a widely used distance-based classifier, the k-nearest neighbors (KNN) technique, is utilized to identify the health status of any new symbol sequence. In addition to the KNN methods, the decision tree (DT) model is also used here for two reasons: (1) it can handle both real-valued and categorical features, which makes it suitable for both raw data and symbolized data; (2) its output results are easy to interpret and explain, which facilitates interactions with the expert team [8].

4.5.2 Vector-Space-Model-Based Method

Since the results of traditional KNN techniques are hard to interpret while the computational cost of decision tree methods is expensive, a recently proposed approach based on a vector space model (VSM) of symbol sequences can be utilized here to provide efficient and interpretable classification [9]. The key idea of VSM is that instead of directly classifying symbol sequences, it first builds class-characteristic weight vectors based on the frequency values of different words, and then utilizes cosine similarity to identify classes. Figure 4.13 illustrates how such a VSM-based classification method works. The detailed procedure can be summarized as follows:

1. Build a bag of words (BOW): The objective of this step is to build a word library from extracted symbol sequences. This can be achieved via first sliding a window of length w across each symbol sequence to extract a list of subsequences. Each unique subsequence is considered as a symbol word, and all of these

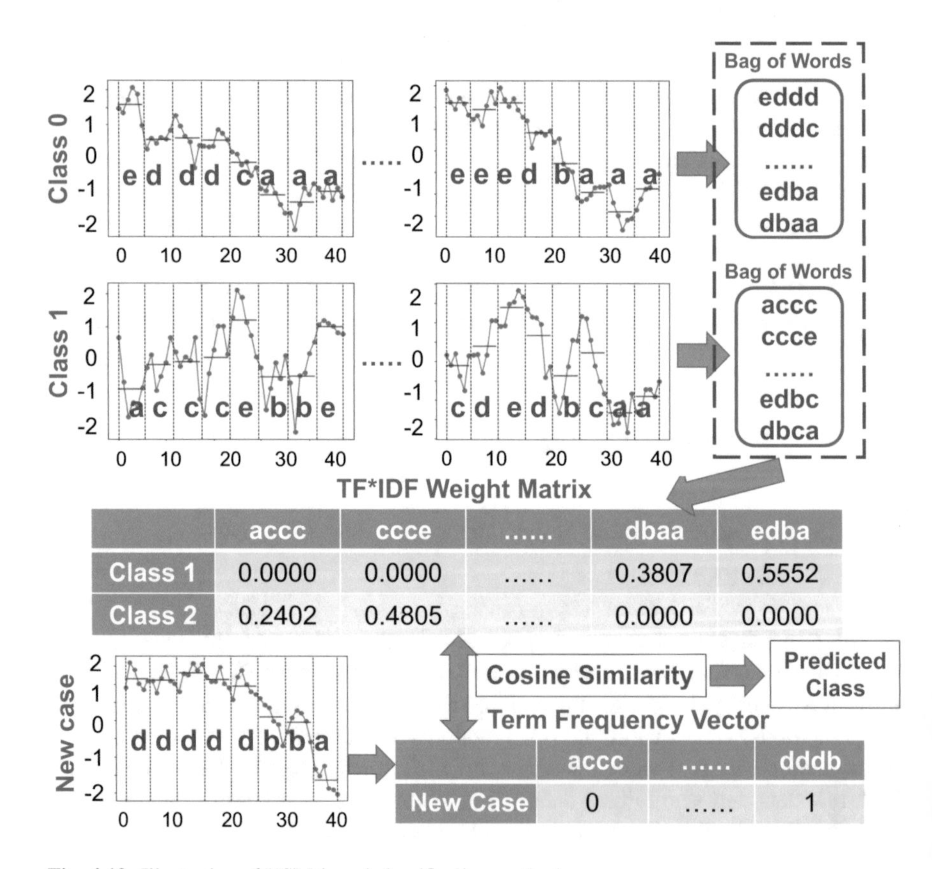

Fig. 4.13 Illustration of VSM-based classification method

subsequences together form the bag of words for the symbol sequences extracted during time series symbolization. For example, as shown in Fig. 4.13, after the raw time-series data have been transformed into SAX symbol sequences, a sliding window of length 4 is used to extract unique words from these symbol sequences, generating a BOW representation in the end.

2. Build vector space model (VSM): Since the vector space model is widely used in the field of information retrieval [10], some terminology are first introduced here: (1) term t: a single word; (2) corpus c: a set of bags; (3) weight matrix M: weights of all words in a corpus. The coefficients in this matrix can be calculated using $TF * IDF$ weighting. Specifically, TF refers to the term frequency and IDF represents the inverse document frequency. They can be computed as follows:

$$TF(t, bow) = \begin{cases} \log(1 + F(t, bow)), & \text{if} \quad F(t, bow) > 0 \\ 0, & \text{otherwise} \end{cases} \tag{4.14}$$

$$IDF(t, c) = \log \frac{|c|}{DF(bow, t)} \tag{4.15}$$

where $F(t, bow)$ represents the frequency of the term t in the bag of words bow and $DF(bow, t)$ represents the frequency of bags of words in the corpus c that contain the term t. Note that $|c|$ is the size of the corpus c. Therefore, the $TF * IDF$ weights can be obtained as:

$$TF * IDF = \log(1 + F(t, bow)) \times \log \frac{|c|}{DF(bow, t)} \tag{4.16}$$

For example, as shown in Fig. 4.13, the bags of words for each class are used to compute the $TF * IDF$ weight matrix M, where each row represents a class' term weight vectors and each column represents weight coefficients for a specific term.

3. Classify new cases: After the term weight matrix is built, the class of unlabeled new symbol sequence can be identified. First, the term frequency vector for the new case v is calculated using the same sliding window and bags of words. Next, the Cosine similarity between this term frequency vector v and each row vector in the term weight matrix M is calculated as:

$$\cos(v, M_i) = \frac{v \cdot M_i}{||v|| \cdot ||M_i||} \tag{4.17}$$

where M_i refers to the ith row in M and can be considered as the term weight vector for class i. Finally, the label of the new case is obtained by finding the row vector that outputs the highest Cosine similarity:

$$label(v) = \arg\max_i \cos(v, M_i) \tag{4.18}$$

For example, as shown in Fig. 4.13, after the new time series has been transformed into a new SAX symbol sequence, the same sliding window and bag of words are used to compute the term frequency vector. The final label for this new case is thus determined by the class that has highest Cosine similarity.

4.6 Experiments and Results

We carried out experiments using the "NE40E" core router product. The details of this core router have been described in previous chapters. Twenty "NE40E" core routers with similar software platform version and hardware configuration were monitored in real time by a distributed agent-based system. A total of 2892 features were extracted and sampled every 30 min for 90 days of operation of each "NE40E" core router, generating a set of twenty multivariate time-series data consisting of 4320 time points. In our experiments, we deliberately change version, configuration, and running tasks on core routers so that we can cover as many different system states as possible in order to make our model as comprehensive as possible. Besides the fourfold cross validation, a set of metrics is also needed to effectively assess the quality of the proposed health analyzer.

4.6.1 Metrics for Health Analysis

The health status of commercial core routers is too complex to be precisely determined because the number of intermediate running states of core routers can become extremely large as time proceeds, e.g., about 2000–4000 states for the NE40E over 90 days of operations. Therefore, after interacting with the expert team, we defined six health levels to represent the overall health status of our experimental core routers in a meaningful way, as shown in Fig. 4.14:

1. Health Level 0: Assume that a g-size normal global pattern set $GP = \{gp_0, gp_1, \ldots, gp_g\}$ and a l-size normal local pattern set $LP = \{lp_0, lp_1, \ldots, lp_l\}$ have been learned from the proposed symbol-based pattern-learning component. A time-series instance D is labeled as "Health Level 0" if and only if its global pattern $D_{gp} \in GP$ and local pattern set $D_{LP} \subseteq LP$. Therefore, "Health Level 0" indicates that the monitored core router is running in a healthy manner without any obvious abnormal operations.
2. Health Level 1: If the global pattern D_{gp} of a time-series instance D can find a perfect match in the g-size normal global pattern set GP, but a small number of elements in the local pattern set D_{LP} of D lie outside the l-size normal local pattern set LP, this time-series instance is defined as "Health Level 1". Core routers having this health status are considered healthy with some minor suspect local patterns. An example of a "Health Level 1" time-series instance is shown

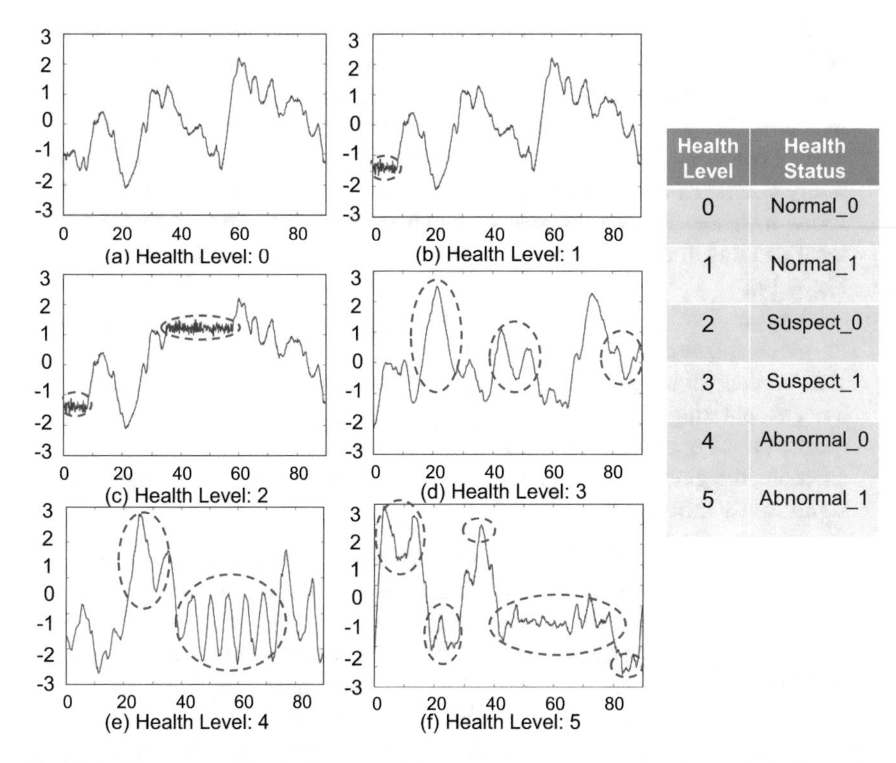

Health Level	Health Status
0	Normal_0
1	Normal_1
2	Suspect_0
3	Suspect_1
4	Abnormal_0
5	Abnormal_1

Fig. 4.14 Illustration of health levels used in our experiments. (**a**) Health Level: 0. (**b**) Health Level: 1. (**c**) Health Level: 2. (**d**) Health Level: 3. (**e**) Health Level: 4. (**f**) Health Level: 5

in Fig. 4.14b. We can see that this time-series instance is almost the same as the "Health Level 0" time-series instance shown in Fig. 4.14a, and the only difference is the minor local pattern at the beginning of the time-series instance.

3. Health Level 2: A time-series instance is identified as "Health Level 2" if its global pattern $D_{gp} \in GP$ while its local pattern set D_{LP} is not significantly overlapped with the l-size normal local pattern set LP. This health level indicates that the monitored core router is in fairly good condition over the long term but may encounter anomalies during some time periods. Figure 4.14c shows an example of a "Health Level 2" time-series instance. It can be seen that a number of local patterns in the middle of this time-series instance is significantly different from that in Fig. 4.14a.
4. Health Level 3: If a time-series instance D can only find a partial match for its global pattern D_{gp} in the g-size normal global pattern set GP, it is classified as "Health Level 3". Such a distortion of global patterns is caused by either true system anomalies or just external noise. In this case, the monitored core router is in a suspect unhealthy warning state. From Fig. 4.14d, we can see that the overall trend of this time-series instance is similar to that in Fig. 4.14a, but the general shape is distorted.

5. Health Level 4: If a large number of elements in the local pattern set D_{LP} of a time-series instance D lie outside of the l-size normal local pattern set LP, this time-series instance is defined as "Health Level 4". This health level indicates that even though the monitored core router can still execute some normal operations, its performance and efficiency are severely affected by critical faulty components. Figure 4.14e shows an example of "Health Level 4" time-series instance. We can see that although the overall trend is preserved, most local patterns in this time-series instance are different from the local patterns in Fig. 4.14a
6. Health Level 5: A time-series instance D is labeled as "Health Level 5" if its global patterns D_{gp} is significantly different from all elements in the g-size normal global pattern set GP. Core routers staying in this health level are encountering severe health problems that prevent them from continuing most normal operations and may even lead to a complete system crash. For example, the global shape and trend of the time-series instance in Fig. 4.14f are significantly different from those in Fig. 4.14a.

After defining the health level of our experimental core routers, we introduce two metrics (diagnosis accuracy and storage required) to evaluate the overall performance of the proposed health-analysis system. The diagnosis accuracy, referred to as a percentage, is the ratio of the number of correctly classified health levels to the total number of instances in the testing set. For example, if the testing set has 10 instances, a diagnosis accuracy of 70% means that the health level of 7 out of 10 instances are correctly classified. The storage consumption is the total storage space required to process a set of time-series instances.

In addition, we use fine-grained information-theoretic metrics, called *precision* and *recall*, to comprehensively evaluate the health-analysis system. Positive predictive value (PPV), also known as *precision*, is the proportion of the predicted positive cases that are correct, and it is calculated using (4.19):

$$PPV = \frac{TP}{TP + FP} \tag{4.19}$$

True positive rate (TPR), also known as *recall*, is the proportion of positive cases that are correctly identified, calculated as follows:

$$TPR = \frac{TP}{TP + FN} \tag{4.20}$$

where TP is the number of correctly predicted positive cases; FP is the number of incorrectly predicted positive cases; FN is the number of incorrectly predicted negative cases; TN is the number of correctly predicted negative cases. Since six health levels $H = \{h_0, h_1, \ldots, h_5\}$ are defined in our experiments, PPV_i describes the percentage of instances labeled with health level h_i are successfully classified as h_i, while TPR_i reflects the percentage of instances classified as h_i are truly labeled with health level h_i. A combination of these two metrics provides a comprehensive evaluation of the proposed health-evaluation solution.

4.6.2 Results on Health Analysis

To evaluate the effectiveness of the proposed health analyzer, we have implemented two raw-data-clustering-based baseline algorithms, namely "raw-hac-1nn" and "raw-hac-dt". The first step for both baseline methods is to form a set of clusters $C = \{c_1, \ldots, c_g\}$ using the hierarchical agglomerative clustering (HAC) algorithm. The two baseline methods then utilize a nearest-neighbor (1NN) approach ("raw-hac-1nn") and a decision tree (DT) approach ("raw-hac-dt"), respectively, to classify new time-series instances and evaluate their health level. For the proposed symbol-based health analyzer, the raw time-series data is first symbolized using three symbolization approaches: SAX, 1d-SAX, and NSAR. The next step is to form a number of normal global patterns $GP = \{gp_1, \ldots, gp_g\}$ using the HAC algorithm and a number of normal local patterns $LP = \{lp_0, lp_1, \ldots, lp_l\}$ using the sequitur rule discovery method. Finally, three classification methods: 1NN, DT, and VSM are applied to classify new symbolized time-series instances. Therefore, a total number of 9 combinations are evaluated in our experiments, namely "sax-hac-1nn", "sax-hac-dt", "sax-hac-vsm", "1d-sax-hac-1nn", "1d-sax-hac-dt", "1d-sax-hac-vsm", "nsar-hac-1nn", "nsar-hac-dt", "nsar-hac-vsm".

The first experiment we conduct is to evaluate the overall performance of the proposed health analyzer with increasing time-series length. Figures 4.15

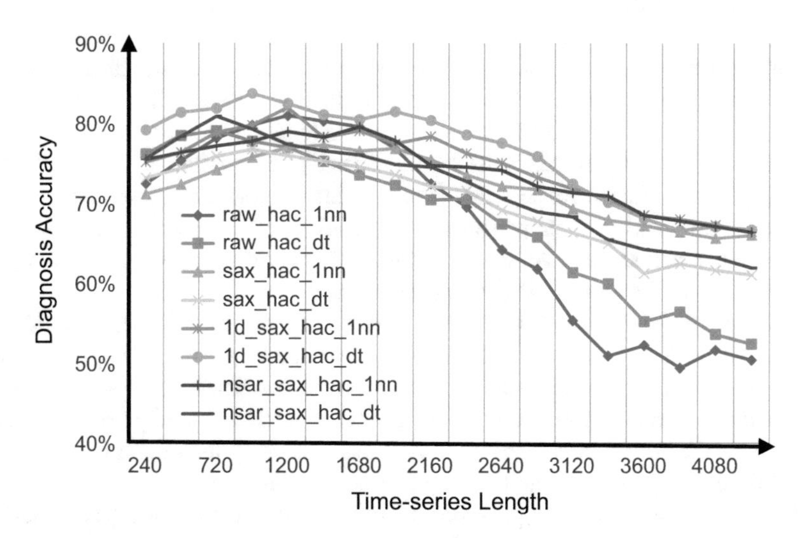

Fig. 4.15 Diagnosis accuracy of health level with increasing time-series length using different methods

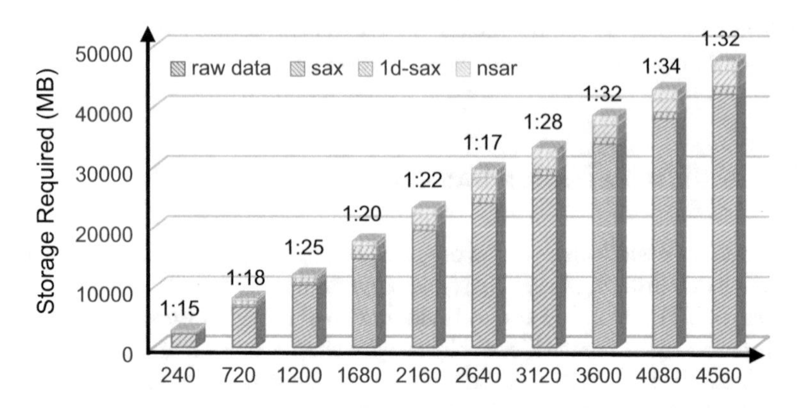

Fig. 4.16 Evaluation of data symbolization in terms of storage requirement as a function of time-series length

and 4.16 show the diagnosis accuracy and storage required for different health analysis methods with increasing time-series length. Note that for the symbol-based methods, since the symbolization of the original time series depends on a number of tunable parameters such as word size and alphabet size, a combination of parameters that gives better overall diagnosis accuracy is selected for each time-series length. The results can be summarized as follows:

1. For the two raw-data-based methods, namely "raw-hac-1nn" and "raw-hac-dt", the overall diagnosis accuracy of "raw-hac-1nn" method increases first but then decreases rapidly as the time-series length increases. We can see from Fig. 4.15 that its diagnosis accuracy gradually increases from 72% to 81% and then drops to 50%. Similarly, although "raw-hac-dt" has a higher diagnosis accuracy for short time series, its diagnosis accuracy also decreases significantly as the time-series length increases. For example, in Fig. 4.15, its diagnosis accuracy decreases from 76% to 52% as the time-series length increases.
2. For the two sax-based methods: "sax-hac-1nn" and "sax-hac-dt", although their diagnosis accuracies are slightly lower than the two raw-data-based methods for shorter time series, their performance degradation with increasing time-series length is much smaller. As shown in Fig. 4.15, their diagnosis accuracy changes from 72% to 66% and 73% to 61%, respectively.
3. For the two 1d-sax-based methods: "1d-sax-hac-1nn" and "1d-sax-hac-dt", we can see from Fig. 4.15 that they have achieved better performance than almost all other methods for both short and long time series. Their diagnosis accuracies remain at around 70% even for the longest time series in our experiment. This is because the 1d-sax approach can improve diagnosis by incorporating additional local trend information into the original sax representation.
4. For the two nsar-based approaches, namely "nsar-hac-1nn" and "nsar-hac-dt", the diagnosis accuracy of "nsar-hac-1nn" is comparable to that of "1d-sax-hac-1nn" and remains high with increasing time-series length. Meanwhile, the

diagnosis accuracy of "nsar-hac-dt" drops a little faster, but still remains much higher than the raw-data-based methods.
5. From Fig. 4.16, we see that the storage requirement (memory) of raw-data-based methods increases nearly linearly from 2200 MB to 42,200 MB as the time-series length increases. In contrast, for the sax-based methods, the storage requirement initially starts at 146 MB and reaches only 1319 MB for the longest time series. Therefore, the compression ratio of the sax-based methods to the raw-data-based methods ranges from 1:15 to 1:34 for different time-series lengths. For the 1d-sax-based methods, although the additional symbols that encode local trend information can increase diagnosis accuracy, they also introduce extra storage cost, achieving compression ratio that is almost half of the sax-based methods. In contrast, the nfar-based approaches are able to compress the raw time series almost as efficient as the sax-based methods.

In summary, when the time-series length increases, the diagnosis accuracy of the two raw-data-based approaches decreases significantly while the reduction in diagnosis accuracy of other symbol-based methods is much less. This is because an efficient symbolization approach can avoid the "curse of dimensionality" that is ubiquitous in data analytics; symbolization effectively maps a high-dimensional time series into a low-dimensional discretized sequence while still preserving critical global and local patterns for diagnosing the health status. Moreover, since the time cost of most machine-learning algorithms is proportional to the amount of data needed and the symbolization procedure itself does not cost much time, the computational time of the proposed method is also less than the traditional approaches.

After evaluating the overall diagnosis accuracy and storage consumption of the proposed health analyzer, a second experiment was conducted to assess the effectiveness of the health analyzer on each individual health level. Figures 4.17, 4.18, 4.19, and 4.20 show the precision and recall for the six health-level classes using the four different health-analysis approaches. The results can be summarized as follows:

1. Both the two raw-data-based baseline methods and all the other symbol-based methods achieve relatively high precision and recall for classifying "Health Level 0". This is because only historical time-series instances, whose global and local patterns are all normal, are labeled as "Health Level 0" during the training phase, leaving little room for variations. Therefore, all the eleven methods can effectively classify a new "Health Level 0" time-series instance as long as a sufficient number of normal patterns have been learned.
2. The two raw-data-based methods perform better than the two sax-based methods in classifying "Health Level 1". The reason is that "Health Level 1" time-series instances only contain a small number of deviated local patterns. The proposed sax-based methods may not be able to accurately capture these minor outliers because information about some short-length subsequences are lost during discretization and compression. In contrast, both the 1d-sax-based approaches and nsar-based methods achieve comparable or even better precision and recall

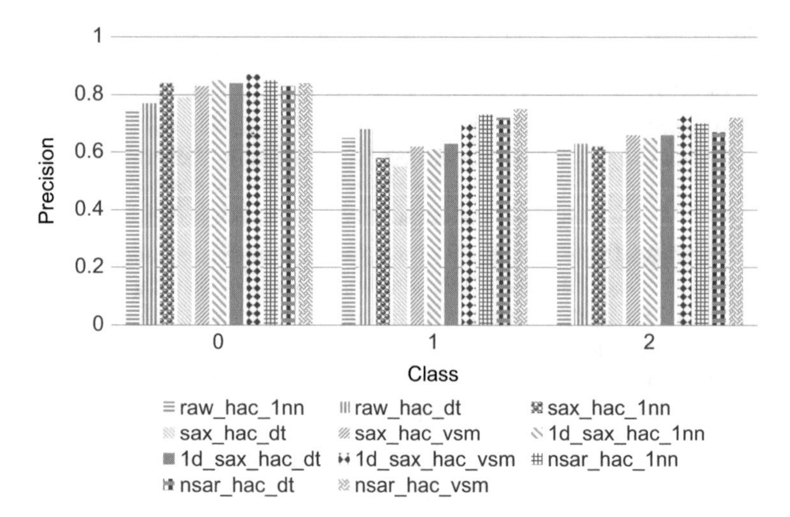

Fig. 4.17 Precision of health level 0–2 using different methods

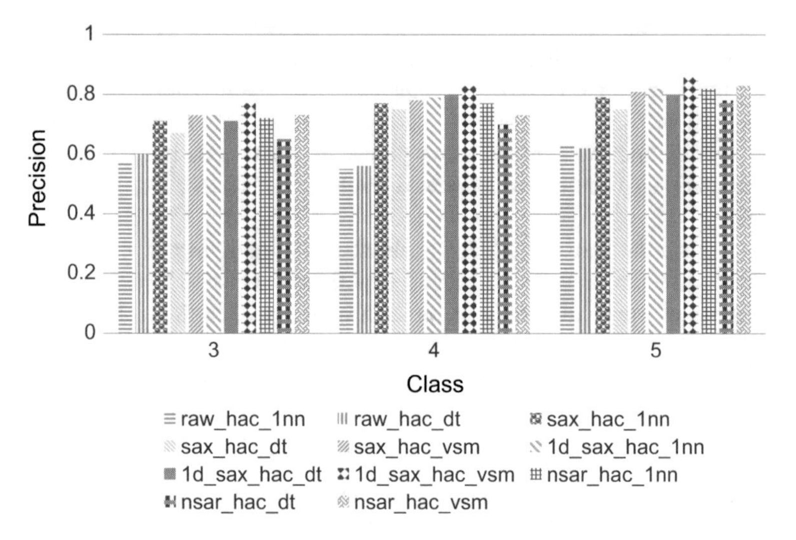

Fig. 4.18 Precision of health level 3–5 using different methods

than the baselines in classifying "Health Level 1". This is because the 1d-sax symbolization retains local trend information while the nsar technique extracts multi-scale characteristics.

3. For the class "Health Level 2", similar values of precision and recall (around 0.6) are achieved by the raw-data-based methods and the sax-based methods. One possible reason is that a number of anomalous local patterns occur in "Health Level 2" time-series instances. Whether such types of anomalous local patterns can be identified depends on the segmentation of time-series instances for both

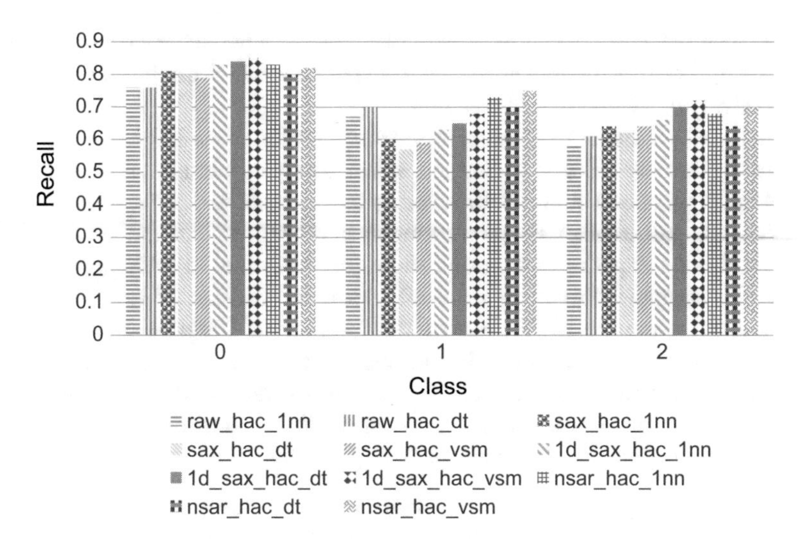

Fig. 4.19 Recall of health level 0–2 using different methods

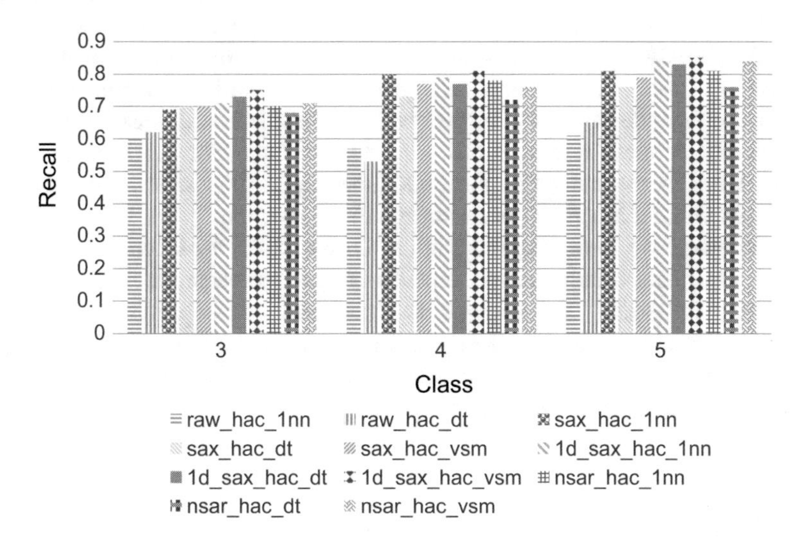

Fig. 4.20 Recall of health level 3–5 using different methods

raw-data-based and symbol-based methods. In contrast, the methods "1d-sax-hac-vsm" and "nsar-sax-hac-vsm" perform much better than other approaches. One possible explanation is that the vector space model built from symbol sequence can efficiently identify critical local patterns.

4. For the remaining three classes: "Health Level 3", "Health Level 4", and "Health Level 5", the nine symbol-based approaches all achieve much higher precision and recall than the two raw-data-based approaches. This is because time-series

instances belonging to these three health classes have either heavily distorted shape or a large number of abnormal local patterns, which can be identified more effectively by symbol-based methods.

5. Among all methods, the "1d-sax-hac-vsm" and "nsar-sax-hac-vsm" have consistently achieved better performance in classifying almost all health levels. This phenomenon could be attributed to two factors: one is the symbolization procedure and another is the classification procedure. For the symbolization methods, the experimental results show that both the 1d-SAX approach and the NSAR approach can retain useful local information of raw time series. Moreover, since the NSAR can capture multi-scale irregular or hidden local patterns, it performs better in identifying the two suspect health-status levels. For the classification procedure, the experiments indicate that the vector space model can learn decision boundary from symbol sequences more efficiently, giving higher diagnosis accuracy and interpretability.

In summary, for the long-term time-series instances extracted in our experiment, the two raw-data-based baseline algorithms perform well for the two normal health classes "Health Level 0" and "Health Level 1". However, they do not achieve satisfactory precision and recall levels for the two suspect health classes "Health Level 2", "Health Level 3" and the two abnormal health classes "Health Level 4", "Health Level 5". In contrast, the proposed symbol-based methods perform much better for classifying the suspect and abnormal health classes. One possible explanation for these results is that a number of anomalous local patterns and distorted shapes are hidden in the high-dimensional time series and cannot be discovered without proper compression and transformation.

In summary, we have found through extensive experimentation that the effectiveness of system health analysis depends on the representation of the extracted time-series instances. The proposed symbol-based health analyzer utilizes the advantages of clustering and classifying discretized symbol sequence, leading to higher diagnosis accuracy and lower storage requirement for identifying the health status of complex core router systems.

4.7 Conclusion

We have described the design of a symbol-based health analyzer for a complex core router system. Various symbol-representation methods have been implemented to compress long-term time-series in a hierarchical way without losing important local information. Both hierarchical agglomerative clustering and a sequitur rule discovery have been implemented to learn global and local patterns. Three classification methods have also been implemented to identify the health status of core routers. The effectiveness of the proposed health analyzer has been validated by data collected from a set of commercial core routers. Experimental results show that the proposed symbol-based health analyzer is significantly more effective and efficient in ascertaining health status compared to baseline methods that utilize raw data.

References

1. J. Lin, E. Keogh, L. Wei, S. Lonardi, Experiencing sax: a novel symbolic representation of time series. Data Min. Knowl. Disc. **15**, 107–144 (2007)
2. S. Malinowski, T. Guyet, R. Quiniou, R. Tavenard, 1d-sax: a novel symbolic representation for time series, in *International Symposium on Intelligent Data Analysis* (2013), pp. 273–284
3. E. Grafarend, *Linear and Nonlinear Models: Fixed Effects, Random Effects, and Mixed Models* (Walter de Gruyter, New York, 2006)
4. X. He, C. Shao, Y. Xiong, A non-parametric symbolic approximate representation for long time series. Pattern Anal. Appl. 19, 111–127 (2016)
5. O. Maimon, L. Rokach, *Data Mining and Knowledge Discovery Handbook* (Springer, Berlin, 2005)
6. C.G. Nevill-Manning, I.H. Witten, Identifying hierarchical structure in sequences: a linear-time algorithm. J. Artif. Int. Res. **7**, 67–82 (1997)
7. Y. Liao, V.R. Vemuri, Use of k-nearest neighbor classifier for intrusion detection. Comput. Secur. **21**, 439–448 (2002)
8. J. Quinlan, Induction of decision trees. Mach. Learn. **1**(1), 81–106 (1986)
9. P. Senin, S. Malinchik, Sax-vsm: interpretable time series classification using sax and vector space model, in *IEEE International Conference on Data Mining (ICDM)* (2013), pp. 1175–1180
10. G. Salton, A. Wong, C.-S. Yang, A vector space model for automatic indexing. Commun. ACM **18**, 613–620 (1975)

Chapter 5
Self-learning and Efficient Health-Status Analysis

Although a large amount operational data is collected from core routers, due to high computational complexity and expensive labor cost, only a small part of this data is labeled by experts. The lack of labels is an impediment towards the adoption of supervised learning. Therefore, in this chapter, we present an iterative self-learning procedure for assessing the health status of a core router. This procedure first computes a representative feature matrix to capture different characteristics of time-series data. Not only statistical-modeling-based features are computed from three general categories, but also a recurrent neural network-based autoencoder is utilized to capture a wider range of hidden patterns. Moreover, both minimum-redundancy-maximum-relevance (mRMR) method and fully-connected feedforward autoencoder are applied to further reduce dimensionality of extracted feature matrix. Hierarchical clustering is then utilized to infer labels for the unlabeled dataset. Finally, a classifier is built and iteratively updated using both labeled and unlabeled dataset. Field data collected from a set of commercial core routers are used to experimentally validate the proposed health-status analyzer. The experimental results show that the proposed feature-based self-learning health analyzer achieves higher precision and recall than the traditional supervised health analyzer as well the currently deployed rule-based health analyzer. Moreover, it achieves better performance than the three anomaly detection baseline methods under the transformed binary classification scenario.

The remainder of the chapter is organized as follows. Section 5.1 presents the motivation of self-learning health analysis for core routers. Section 5.2 describes the framework of feature-based self-learning health analyzer. Sections 5.3 and 5.4 present how a wide range of features are calculated from multivariate time series and how different health statuses are identified from partially-labeled feature matrix. Experimental results for a commercial core router system are presented in Sect. 5.5. Finally, Sect. 5.6 concludes the chapter.

S. Jin et al., *Anomaly-Detection and Health-Analysis Techniques for Core Router Systems*, https://doi.org/10.1007/978-3-030-33664-6_5

5.1 Motivation

The design of a health-status analyzer is more difficult than the implementation of an anomaly detector because:

1. Anomaly detection is unsupervised while health analyzer requires fully-labeled data. However, the volume of operational data collected from commercial core routers can reach TB levels, making it infeasible for experts to label the data manually;
2. Classifying complex time-series data is harder than detecting anomalous time-series data because subtle differences between a pair of time series must also be identified by the classifier.

Although a symbol-based health analyzer for core routers has been presented in Chap. 4, it still requires fully-labeled data during its training phase. Data instances are considered as being labeled only after the expert team has determined their normal/abnormal conditions, which is difficult to obtain in the early stages of monitoring. Moreover, although symbolization can reduce the time cost as well as the storage requirement, some critical local information may be lost during symbolization. Therefore, in this work, we use feature extraction and selection techniques as well as a deep-learning-based autoencoder to characterize complex time series. A self-learning approach is then implemented to analyze the health status of core routers using partially labeled data.

5.2 Framework of Feature-Based Self-learning Health Analysis

The key idea in the proposed method (Fig. 5.1) is that instead of directly analyzing the health status from a large volume of raw time series data, we first extract and select a set of features that capture the characteristics of high-dimensional time series. The notation of a feature in this chapter is different from the definition of a feature in previous Chaps. 2–4, where features refer to the temporal measurements of different monitored items (variables) in core routers. The "features" in this chapter are defined as metrics calculated from the raw time series of the variables, and they represent various local and global characteristics of the time series. The steps involved in our procedure are as follows:

1. Feature extraction and selection: Since each feature is a low-dimensional measurable characteristic of the time series, extracting and selecting a set of representative features provides a more complete understanding of the time series. This component takes a set of clean and aligned time series as input, and outputs a representative feature matrix to the self-learning component.

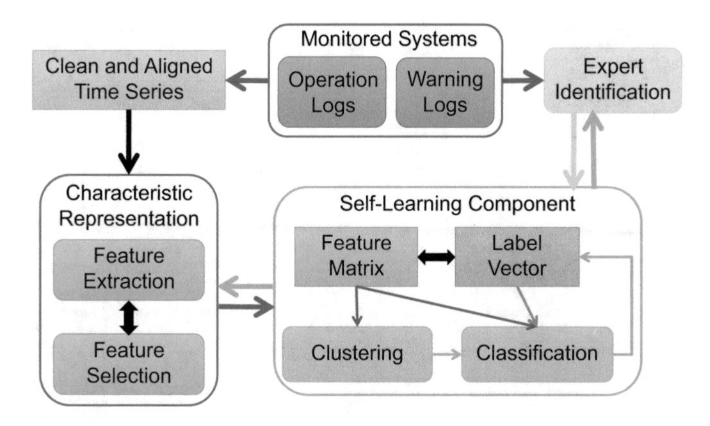

Fig. 5.1 An illustration of the proposed feature-based and self-learning health-status analyzer

2. Expert identification: This component is maintained and updated by an expert team. The experts first label a limited number of time series instances using historical warning logs and their rule tables. This set of labels then serves as the initial label vector to the self-learning component. During the self-learning procedure, newly updated labels are also fed to this component for checking.
3. Self-learning component: This component consists of two parts—clustering and classification. The objective of clustering is to increase the number of labeled instances. Since similar instances are grouped together after clustering, the label value of labeled instances can be propagated to unlabeled instances within the same cluster. The classification part is used to identify the health status of the system by iteratively learning a model from partially labeled data.

5.3 Feature Extraction and Selection

As shown in Fig. 5.2, Assume that m router slots are monitored, where each router slot has n monitored items across the temporal domain. Therefore, a total of $m \times n$ time-series instances are collected in the original dataset: $D = \left\{d_1^{(1)}, d_2^{(1)}, \ldots, d_n^{(1)}, d_1^{(2)}, \ldots, d_n^{(2)}, \ldots \ldots, d_1^{(m)}, \ldots, d_n^{(m)}\right\}$, where $d_j^{(i)}$ represents the time series sequence extracted from the jth monitored item in the router slot i: $d_j^{(i)} = \{t_1, t_2, \ldots, t_v\}$, where v is the number of time points in $d_j^{(i)}$. A set of feature metrics $F = \{F_1, F_2, \ldots, F_u\}$ is then computed using $d_j^{(i)}$ to capture various characteristics of this time-series sequence. Specifically, three types of feature metrics [1] are considered:

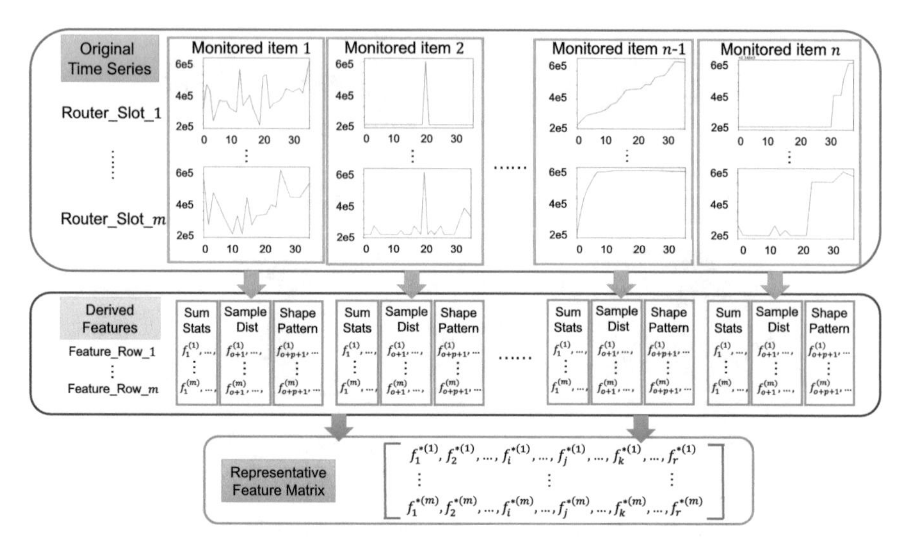

Fig. 5.2 Illustration of feature extraction and selection

1. Summary Statistics: Metrics in this category compute basic summary statistics of time series. For example, the metric F_{mean} computes the arithmetic mean of a time series $d_j^{(i)}$: $F_{mean}(d_j^{(i)}) = \frac{1}{v}\sum_{k=1}^{v} t_k$ and the metric F_{var} calculates the variance of $d_j^{(i)}$: $F_{var}(d_j^{(i)}) = \frac{1}{v}\sum_{k=1}^{v}(t_k - \bar{t})^2$. Other metrics such as F_{max}, F_{min}, F_{std}, F_{skew}, F_{kurt}, F_{median}, etc., are used to obtain the maximum, minimum, standard deviation, skewness, kurtosis, median, etc., of $d_j^{(i)}$, respectively.
2. Sample Distribution: Metrics in this category compute characteristics of a time series using different sample distributions. For example, the metric $F_{entropy}$ first samples a time series $d_j^{(i)}$ into b equidistant bins, and then calculates the entropy as: $\sum_{k=0}^{\min(b,v)} p_k \log(p_k) 1_{(p_k>0)}$, where p_k is the percentage of samples in bin k. Another example is the metric $F_{quantile}$. It computes the relative index r where a pre-specified $q\%$ samples of a time series $d_j^{(i)}$ lie to the left of r: $r = \frac{a}{v}$, where $a = \min\left\{p | 1 \leq p \leq v, \frac{\sum_{k=1}^{p} t_k}{\bar{t}} \geq q\%\right\}$.
3. Shape Pattern: Metrics in this category estimate a wide range of shapes and patterns of a time series. For example, regular periodic patterns of a time series are obtained via the autocorrelation metric $F_{autocorrelation} = \frac{1}{var(t)}\sum_{k=1}^{v-l}(t_k - \bar{t})(t_{k+l} - \bar{t})$, where l is the value of lags across the temporal domain. Similarly, the metric F_{cwt_peaks} detects different peaks (bursty patterns) of a time series. For this purpose, F_{cwt_peaks} first transforms the time series $d_j^{(i)}$ into Ricker wavelets [2] with widths ranging from 1 to l. A peak is then identified if the signal-to-noise ratio (SNR) of a local maximum of wavelets is sufficiently high. Some general shapes/trends of a time series can also be obtained by metrics in this category. For instance, the metric F_{linear_trend} determines a linear least-squares regression fit for a time series so that a set of increasing/decreasing trends can be estimated.

After these feature metrics are applied to all the extracted $m \times n$ time-series instances, an initial $m \times n_u$ feature matrix is formed, where n_u is the number of features obtained from u feature metrics. However, not every feature is useful for health-status analysis. Irrelevant and redundant features may impair the effectiveness of the subsequent self-learning procedure. Therefore, feature selection is needed to select an effective, but reduced, set of features.

Although the minimum-redundancy-maximum-relevance ($mRMR$) metric is widely used for feature selection [3], this method requires fully labeled input data, which is not available during the early stage of our self-learning procedure. We thus extend the $mRMR$ metric so that it can handle partially labeled input data. Suppose we have a set of n_u features $\mathbf{f} = \{f_1, f_2, \ldots, f_{n_u}\}$ and a set of w labels $\mathbf{a} = \{a_1, a_2, \ldots, a_w\}$. Since input data are partially labeled, a special label $a_0 = -1$ is defined for any unlabeled input instances. The measure $ratio_0 = \frac{count(a_0)}{m}$ is defined to indicate the percentage of unlabeled instances among input data. For each label a_i and a given feature f_j, their mutual information $I(a_i, f_j)$ is calculated as:

$$I(a_i, f_j) = \begin{cases} 0, & \text{if } ratio_0 \geq \alpha \\ \int_{a_i} \int_{f_j} p(a_i, f_j) \log \frac{p(a_i, f_j)}{p(a_i)p(f_j)} da_i df_j, & \text{otherwise} \end{cases} \quad (5.1)$$

where $p(a_i)$ and $p(f_j)$ are probability density functions of a_i and f_j, and $p(a_i, f_j)$ is the joint probability density function of a_i and f_j. The parameter α is a predefined threshold used to limit the maximal allowable ratio of unlabeled instances. We calculate the relevance value $D(\mathbf{a}, f_j)$ between the feature f_j and label set $\mathbf{a}$ as follows:

$$D(\mathbf{a}, f_j) = \frac{1}{w} \sum_{a_i \in \mathbf{a}} I(a_i, f_j) \quad (5.2)$$

The MaxRel set $\mathbf{f}' = \{f'_1, f'_2, \ldots, f'_k\}$ is a selected subset of the top k features having the highest relevance values. The set $\mathbf{f}'$ is further evaluated by computing its redundancy value $R(\mathbf{f}')$, as shown below:

$$R(\mathbf{f}') = \frac{1}{|\mathbf{f}'|^2} \sum_{f'_i \in \mathbf{f}'} \sum_{f'_j \in \mathbf{f}'} I(f'_i, f'_j) \quad (5.3)$$

where $I(f'_i, f'_j)$, $f'_i \neq f'_j$, is the mutual information between f'_i and f'_j. We then calculate the minimum-redundancy maximum-relevance ($mRMR$) value as follows:

$$mRMR(\mathbf{f}') = \frac{1}{|\mathbf{f}'|^2} \sum_{f'_k \in \mathbf{f}'} D(\mathbf{a}, f'_k) - R(\mathbf{f}'). \quad (5.4)$$

f_1	f_2	f_3	f_4	f_5	*label*
0.7	0.2	0.4	0.09	0.3	-1
0.8	0.1	0.2	0.25	0.5	1
0.8	0.3	0.6	0.81	0.9	2
0.8	0.4	0.8	0.04	0.2	1

f_2	f_5	*label*
0.2	0.3	-1
0.1	0.5	1
0.3	0.9	2
0.4	0.2	1

Fig. 5.3 An example of $mRMR$-based feature selection

We can next determine the minimum-redundancy-maximum-relevance ($mRMR$) feature subset $\mathbf{f}*$ with the largest $mRMR$ value, as follows: $\mathbf{f}* = \max_{\mathbf{f}'}\{mRMR(\mathbf{f}')\}$.

A simple example of $mRMR$-based feature selection is shown in Fig. 5.3. The original input data consists of four instances, where each instance contains five features and one label (unlabeled instance is denoted as -1). After applying the $mRMR$-based feature selection, the low-relevance features f_1, and the highly redundant features f_3 and f_4 are removed, and the final $mRMR$ feature subset is $\mathbf{f}* = \{f_2, f_5\}$.

Moreover, extended $mRMR$-based feature selection can be easily integrated with the self-learning procedure. In the early stages of self learning, only a limited number of instances are labeled, therefore, $I(a_i, f_j) = 0$ holds true in Eq. (5.1). In this case, the first term in Eq. (5.4) approaches 0 and only the redundancy value $R(\mathbf{f}')$ is important. The feature-selection component thus removes only redundant features and maintain all the other features for the next iteration. As the self-learning procedure continues, more instances are labeled, and the relevance part of $mRMR$ computation becomes important. In this way, both irrelevant and redundant features are eliminated.

5.3.1 Autoencoder-Based Feature Learning

Feature extraction and selection step described above suffers from two limitations. First, we have observed from our experiment that even after the $mRMR$-based feature selection, the number of features is still much larger than the number of available instances, making it difficult for some types of classifiers to be effective. Moreover, some of the previously extracted features are sparse, making it possible to further compress the feature matrix without significant information loss. Second, such a feature extraction step is ad hoc and depends on the experience of experts. Although various characteristics of time series have been extracted, it is hard to ascertain whether the extracted features are sufficient to cover most characteristics of the time-series data. It is possible that some critical characteristics are missed. Therefore, in this paper, the LSTM-based autoencoder is utilized to capture a wider range of hidden patterns. These new approaches are more general and they better match realistic scenarios.

The traditional artificial neural network (ANN) is a supervised machine learning method that is widely used for pattern classification and related problems [4, 5]. The autoencoder is an unsupervised variant of ANN for learning an efficient representation (encoding) of a set of data [6, 7]. As shown in Fig. 5.4, the simplest form of an autoencoder is a three-layer feedforward artificial neural network. It consists of an input layer, an output layer and a hidden layer. Neurons are arranged in layers, and weighted connections link the neurons in different layers. An autoencoder network can be generally divided into two parts: the encoder that compresses the data from input layer into a short code and the decoder that uncompresses that code and produces a reconstruction in the output layer. Therefore, the objective of an autoencoder is to learn a reduced but meaningful representation that can reconstruct the original data as much as possible. The behavior of an autoencoder depends on both the weights (synaptic strength of neuron connections) and the transfer function (input-output function of neurons).

For example, in Fig. 5.4, the number of neurons n in the input layer and output layer are both 5 while the number of neurons m in the hidden layer is 3. First, the value h_i in the hidden layer can be obtained by:

$$h_i = f_{transfer}(\sum_{j=1}^{n} x_j \times w_{ji}) \tag{5.5}$$

where x_j represents the jth neuron in the input layer, w_{ji} represents the weights between the jth neuron in the input layer and the ith neuron in the hidden layer, $f_{transfer}(\cdot)$ represents the transfer function in the ith neuron in the hidden layer. Next, the value y_j in the output layer can be obtained by:

$$y_j = g_{transfer}(\sum_{i=1}^{m} h_i \times w_{ij}) \tag{5.6}$$

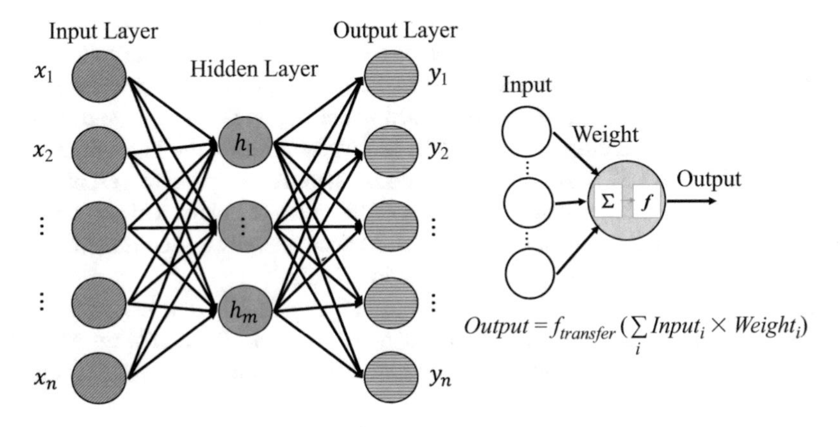

Fig. 5.4 An example of a three-layer autoencoder

where w_{ij} represents the weights between the ith neuron in the hidden layer and the jth neuron in the output layer, $g_{transfer}(\cdot)$ represents the transfer function in the jth neuron in the output layer. The reconstruction error (loss function) can thus be calculated as:

$$\begin{aligned} L &= \sum_{j=1}^{n} |x_j - y_j|^2 = \sum_{j=1}^{n} |x_j - g_{tran}(\sum_{i=1}^{m} h_i \times w_{ij})|^2 \\ &= \sum_{j=1}^{n} |x_j - g_{tran}(\sum_{i=1}^{m} f_{tran}(\sum_{j=1}^{n} x_j \times w_{ji}) \times w_{ij})|^2 \end{aligned} \tag{5.7}$$

The objective of learning an autoencoder is to minimize the above loss function. The back-propagation algorithm is widely used during training phase; it minimizes the loss function of reconstructing original data using gradient descent. When the loss is less than a pre-defined threshold (performance goal) or converges, the training is deemed to be complete. After an autoencoder is built, its encoder part can be used to compress new-incoming data $X = \{x_1, x_2, \ldots, x_n\}$ into a reduced but meaningful representation $H = \{h_1, \ldots, h_m\}$.

We utilize the autoencoder in two different ways in our characteristic representation component. The first method is to apply it to the feature matrix obtained from our previous feature extraction and selection step to further reduce the feature dimension. The second method is to directly apply it to the original multivariate time-series data to learn a set of autoencoder-based features.

Traditional fully-connected feed-forward neural network are not suitable for encoding/decoding time-series data in an autoencoder because of the non-stationary dynamics/patterns within the temporal ordering of the input. Instead, the recurrent neural network (RNN) is promising because it maintains an internal state of the network via a directed cycle of connection between neurons, which allows it to exhibit dynamic temporal behavior [8]. In addition, the hidden state in RNNs is shared over time and thus can contain information from an arbitrarily long window. The Long Short Term Memory (LSTM) serves as the RNN architecture used in autoencoder because it explicitly introduces a memory unit, called the cell, into the network so that long-term historical information can be recalled as needed [9]. As shown in Fig. 5.5, the original feed-forward neural network encoder and decoder are now replaced by multiple LSTM cells. In this framework, the input time-series data are first fed to LSTM encoder cells step-by-step. An encoded representation is then learned from hidden states or outputs of these LSTM cells. This compressed representation is then fed as inputs to the LSTM decoder cells, generating the output time series that closely matches the original input data.

Figure 5.6 shows the zoom-in architecture of a typical LSTM cell. The input, weight, system state and memory cell are denoted as x_t, S_t, W and C_t, respectively. This memory cell keeps or updates information over many time steps and it consists of:

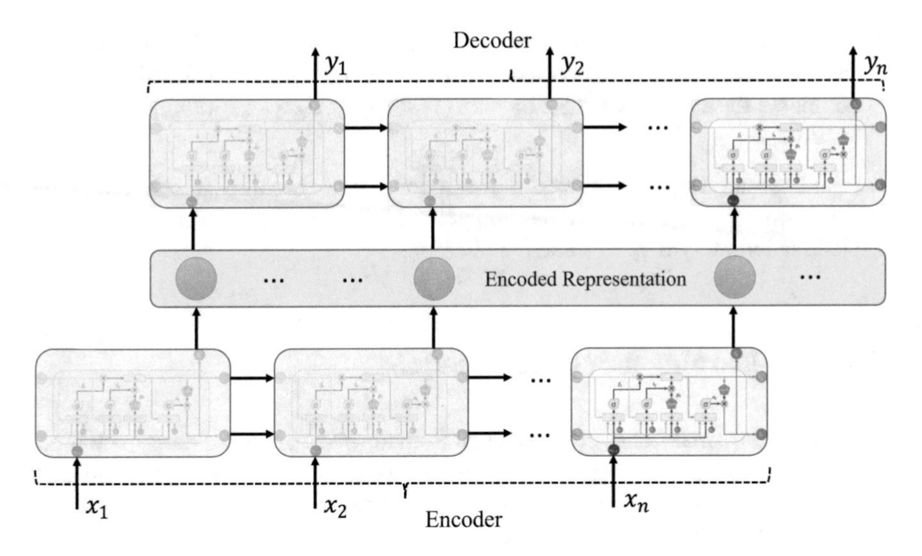

Fig. 5.5 The framework of LSTM-based autoencoder

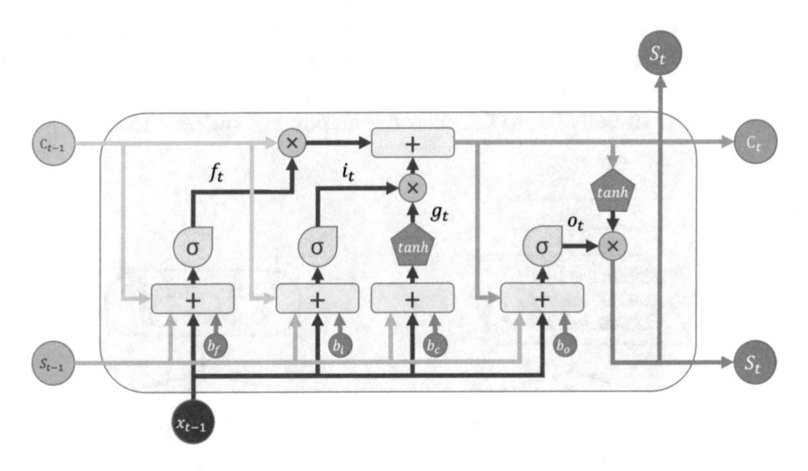

Fig. 5.6 Illustration of the long short term memory (LSTM) method

1. What information will be discarded from the cell state. This decision is made by a sigmoid function σ, called the forget gate layer $f_t = \sigma(W^{(f)}(S_{t-1}, C_{t-1}, x_t) + b_f)$. The layer f_t outputs a number between 0 and 1, where 1 (0) represents "completely keep" ("completely forget").
2. Decide what new information will be stored in the cell state. The decision consists of two parts: a sigmoid function σ, called the input gate layer i_t, decides which values will be updated, and a $tanh$ function g_t that creates a vector of new candidate values: $i_t = \sigma(W^{(i)}(S_{t-1}, C_{t-1}, x_t) + b_i)$; $g_t = tanh(W^{(C)}(S_{t-1}, x_t) + b_C)$. Note that $C_t = f_t \times C_{t-1} + i_t \times g_t$.

3. Decide what will be output from the system. A sigmoid function σ is used to decide what parts of the cell state will be output and a $tanh$ function is used to update the system state S_t: $o_t = \sigma(W^{(o)}(S_{t-1}, C_{t-1}, x_t) + b_o)$; $S_t = o_t \times tanh(C_t)$.

An LSTM can retain information for a long time because it can propagate information without successive multiplication of fractions. The information can be retained indefinitely if $f_t = 1$ and $i_t = 0$.

5.4 Self-learning for Health Analysis

After the feature matrix has been obtained from the original high-dimensional time-series data, it is fed to the self-learning component to train a model for health-status analysis. Since the input data are partially labeled, the learned model and the labeled set are updated iteratively, as shown in Fig. 5.7.

Step 1 Initially, the input data consist of the labeled set $L = L_0$ and the unlabeled set $U = U_0$. The percentage of unlabeled data is then calculated: $r_U = \frac{|U|}{|L|+|U|}$. If r_U is larger than a predefined threshold α, the amount of labeled data is insufficient and a clustering procedure (*Step 2*) is needed to enrich the labeled set. Otherwise, the input data are directly fed to the classifier learning component (*Step 3*)

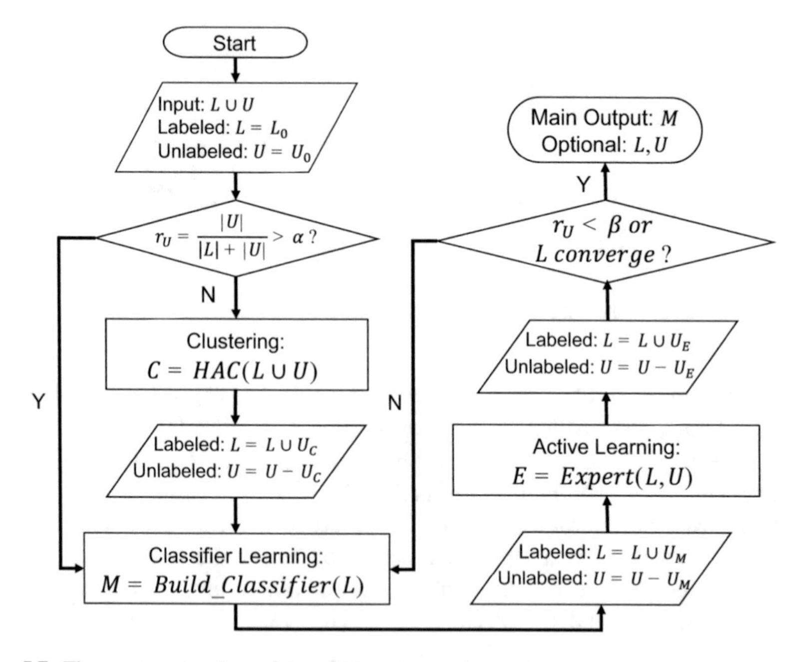

Fig. 5.7 The computation flow of the self-learning component

Step 2 Clustering is performed to propagate labels between similar instances. Specifically, hierarchical agglomerative clustering (HAC) with link constraints [10] is applied to the input data ($L \cup U$), generating a set of clusters $C = \{c_1, c_2, \ldots, c_h\}$. Each cluster c_i can contain both labeled and unlabeled instances. The label value of a labeled instance in c_i is then propagated to its neighboring unlabeled instances in c_i; U_{c_i} is used to denote such a set of unlabeled instances that are now labeled by cluster c_i. A set of unlabeled instances is formed: $U_C = U_{c_1} \cup U_{c_2} \cup \ldots \cup U_{c_h}$. The original labeled and unlabeled sets are thus updated accordingly: $L = L \cup U_C$, $U = U - U_C$.

Step 3 A classifier is learned in this step using the currently available labeled instances L. Machine-learning techniques, such as support vector machine (SVM), decision tree (DT), and artificial neural network, can be used as the base classifier. After training the base classifier with L, a model M is built and applied to the currently unlabeled instances U, which are then classified with different labels $\mathbf{a} = \{a_1, a_2, \ldots, a_w\}$. Besides the label value, a set of class probabilities is also generated from M. The higher the predicted probabilities, the higher the certainty that the unlabeled instances are correctly labeled. A set of unlabeled instances, which are now labeled by classifier M with highest certainty, are formed: $U_M = U_{a_1} \cup U_{a_2} \cup \ldots \cup U_{a_w}$. The labeled and unlabeled sets are thus updated accordingly: $L = L \cup U_M$, $U = U - U_M$.

Step 4 The objective of this step is to interact with the expert identification component. The labeled and unlabeled instances obtained from *Step 3* are fed as queries to the expert team. Experts can answer these queries by (1) rejecting labels assigned to unlabeled instances in *Steps 2–3*; (2) actively adding labels for unlabeled instances; (3) taking no action.

Step 5 Two stopping criteria check whether the self-learning procedure needs another round of iterations. The first criterion is the percentage of current unlabeled data r_U. If r_U is lower than a predefined threshold β, most input data have been labeled and used to update the classifier M. In this case, most information from input data has been utilized and no more iterations are needed. The second criterion is whether the labeled set L approaches convergence. If most elements in the labeled set L remain unchanged for several iterations, it is likely that the learned classifier M has converged. If neither of these two stopping criteria are satisfied, the self-learning procedure will go to *Step 3* for another iteration.

Step 6 After any stopping criterion is met, the proposed self-learning procedure finishes, outputting the up-to-date classifier M as well as the labeled set L and unlabeled set U. Now, when the feature matrix of new time-series instances are extracted, it can be directly fed to the up-to-date classifier M to identify its health status.

We use link-constraint-based hierarchical agglomerative clustering in *Step 2* because:

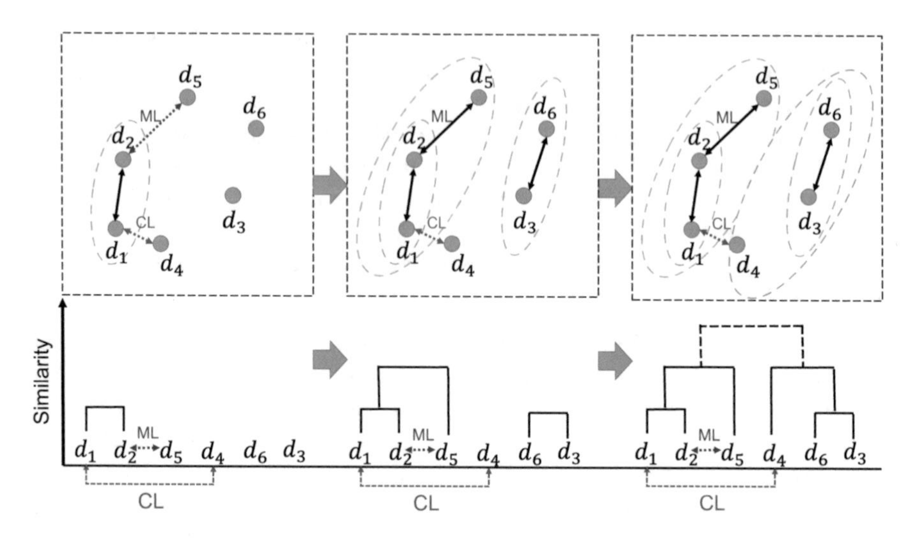

Fig. 5.8 Link-constraint-based hierarchical agglomerative clustering

1. It does not need to specify *a priori* the number of clusters and it outputs a tree diagram, called dendrogram, to illustrate the arrangement of the clusters [11];
2. It allows us to separate/group some data instances by imposing prior link constraints. Since the data used for clustering are partially labeled, we want to ensure that in the labeled set L, instances with different label values will be divided into different clusters while instances with the same label values will be grouped together.

To achieve this, must-link (ML) constraints are imposed on instances with same labels and cannot-link (CL) constraints are imposed on instances with different labels. Figure 5.8 illustrates such link-constraint-based hierarchical agglomerative clustering. The sample dataset D contains six data points, i.e., $D = \{d_1, d_2, d_3, d_4, d_5, d_6\}$. Assume that d_1 and d_4 have different labels while d_2 and d_5 have same label. Therefore, cannot-link is imposed between d_1 and d_4 while must-link is imposed between d_2 and d_5. Although d_1 and d_4 are closest in distance, the cannot-link constraint prevents them from being merged. Instead, d_1 is combined with d_2. Similarly, due to the must-link constraint between d_2 and d_5, d_5 is merged with the cluster (d_1, d_2) instead of d_6.

5.5 Experimental Results

We carried out experiments using the "NE40E" core router product. The details of this core router have been described in previous chapters. The feature extraction component is then applied to the collected data to extract a wide range of

Table 5.1 Features extracted using different feature metrics

Category	Number of metrics	Representative metrics	Number of features
Summary statistics	10	Mean/median	10
		Max/min	
Sample distribution	13	Entropy/quantile	56
		Absolute energy	
Shape pattern	36	Linear trend/cwt peaks	557
		Spectral density	

characteristics. The information regarding features extracted from univariate time series is shown in Table 5.1. A total of 623 features are extracted for each univariate time-series instance. Since each instance in 450 time series has 10 variables, a 450×6230 raw feature matrix is formed after feature extraction.

Meanwhile, as mentioned in Sect. 5.3.1, a LSTM-based autoencoder is also developed to learn a compressed but meaningful representation from the 450 multivariate raw time-series data. The input time-series data can be encoded in both the temporal domain (a total of 2880 time points) and variable space (a total of 10 monitored items), and the performance depends on both the depth and width of the hidden layers in the autoencoder network. After tuning the different parameters, the length of our LSTM-based encoded vector is set to 172 and thus a 450×172 feature matrix is learned after the training of LSTM-based autoencoder.

Since the number of intermediate running states of core routers can be over 1000 for 60 days of operation [12, 13], it is difficult to precisely determine the health status of commercial core routers. Therefore, we combine the pattern library obtained in Chap. 4 with an expert rule table to define categorical health status for experimental core routers. The expert rule table specifies a set of rules for identifying anomalies, where each rule takes the form "IF $\{$variable $m_i = v$ at time $j\}$, THEN $\{$anomaly $a_p = w_j\}$". The pattern library stores a set of normal/abnormal patterns, where each pattern is a symbolized subsequence that characterizes normal/abnormal shape or trend of time series. For each time-series instance collected in our experiment, two counters, namely the anomaly counter and the health counter, are maintained. As long as a matched rule or abnormal pattern is found, its anomaly counter value is increased. In contrast, if a matched normal pattern is found, its health counter value is increased. The final categorical health-status label of this instance is determined by the values of these two counters. If the values of these two counters are both zero, this instance will not be assigned any label.

Without any loss of generality, six labels were defined in our work to represent the overall health status of experimental core routers: (1) Class 0: the system is running in a healthy manner without any obvious abnormal operations (8 out of 450 instances); (2) Class 1: the system is running normally with some minor suspect characteristics (7 instances); (3) Class 2: the system is in relatively good condition with some anomalies (10 instances); (4) Class 3: the system is in a suspect unhealthy warning state (8 instances); (5) Class 4: the system's performance and efficiency are severely affected by critical faulty components (11 instances); (6) Class 5: the system is encountering severe health problems that prevent it from continuing most normal operations (13 instances). The remaining 393 instances are all unlabeled data. An example of these six health labels used in our experiments is shown in Fig. 4.14 in Chap. 4. We can see that the number of matched abnormal patterns increases while the number of matched normal patterns decreases from "Health Level 0" to "Health Level 5". Note that with the improvement of experts' experience, a larger number of categorical labels or even continuous metric values can be used in the future to define the overall system health status in a more comprehensive way.

The information-theoretic metrics *precision* and *recall* are also used here. Their definitions have been introduced in Chap. 4. Since six health levels $H = \{h_0, h_1, \ldots, h_5\}$ are defined in our experiments, the parameter $precision_i$ is the percentage of instances labeled with h_i that are correctly classified as h_i; $recall_i$ reflects the percentage of instances classified as h_i that are truly labeled with h_i. The metric F_1 score uses both *precision* and *recall* to evaluate the effectiveness of different methods. Specifically, the F_1 score is the harmonic mean of *precision* and *recall*, as shown below:

$$F_1 = 2 \times \frac{precision \times recall}{precision + recall} \tag{5.8}$$

5.5.1 Results on Health Analysis

We first evaluate the performance of the proposed health analyzer as a function of the number of iterations. Both SVM and DT have been used as the base classifier in the self-learning procedure, namely "self-svm" and "self-dt". We use SVM and DT as the base classifiers because they are suited for the case where the sample size is relatively small. Another advantage of the DT method is that its output is easy to interpret and for carrying out further root cause analysis. An autoencoder-based self-learning approach, namely "self-autoencoder", is also investigated. In this method, a compressed representation of the time series is learned by an LSTM-based autoencoder and then fed to a fully-connected ANN classifier in the self-learning procedure. Figure 5.9a, b show the F_1 score and the size of labeled set as we increase the number of self-learning iterations. The results can be summarized as follows:

1. For "self-svm", since only 57 out of 450 instances are labeled in the beginning, the initial F_1 score is only 0.47. With self-learning, both the size of labeled set and the corresponding F_1 score increase significantly during the first few iterations. This is because useful information from unlabeled data is rapidly absorbed by the SVM model through self-learning. Both the size of the labeled set and F_1 score converge after tens of iterations, indicating the completion of

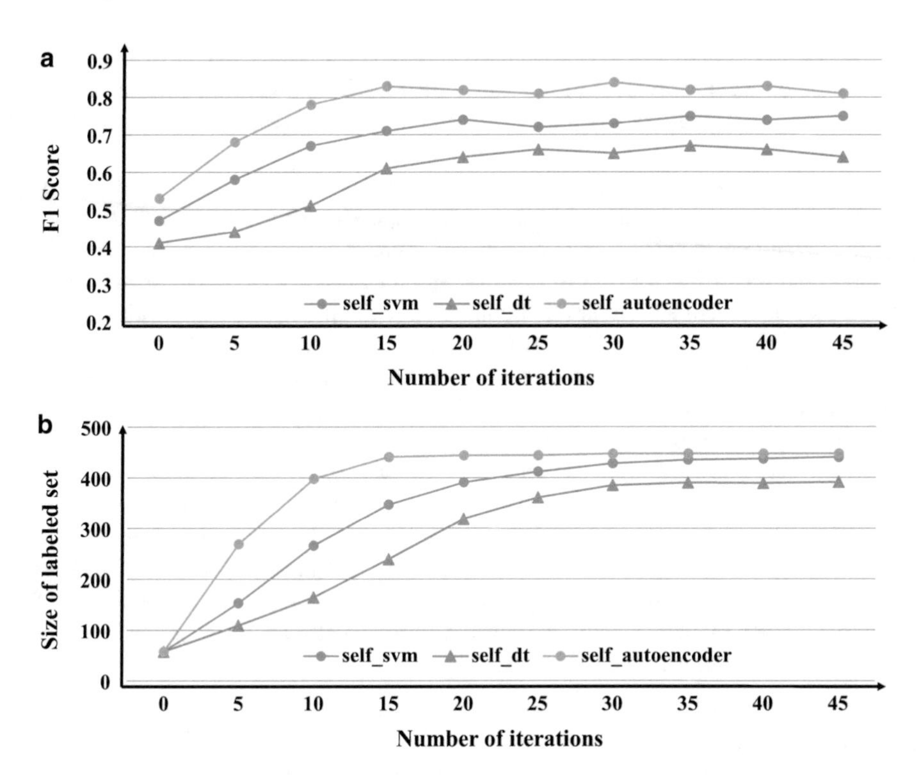

Fig. 5.9 (**a**) F_1 score and (**b**) size of labeled set with an increase in the number of iterations

the self-learning procedure. Eventually, 440 out of 450 instances are labeled and the F_1 score of the up-to-date model is nearly 0.75, which is 60% higher than the initial score.

2. For "self-dt", the initial F_1 score is 0.41. After self-learning, the "self-dt" method labeled 391 instances and achieved an F_1 score of 0.65. These results are worse than that for "self-svm". Moreover, the "self-dt" method converges slower than the "self-svm" method, which indicates that the DT model alone cannot utilize new labeled data as efficiently as the "self-svm" method.
3. For "self-autoencoder", the initial F_1 score is 0.53, which is higher than the corresponding score for both "self-svm" and "self-dt". This is because a more efficient feature representation obtained from autoencoder can help reduce the noise introduced by an insufficient number of labeled instances. Moreover, the "self-autoencoder" method converges faster than the other two approaches and achieves a higher final F_1 score of 0.81. Also, 447 out of 450 instances are successfully labeled in the final stage. The reason that the "self-autoencoder" performs better is because features learned by the LSTM autoencoder implicitly cover more hidden critical "health" information from extracted time-series data, which improves the effectiveness of the self-learning procedure.

A second experiment was conducted to assess the effectiveness of the health analyzer on each individual health class. Five baseline algorithms, namely "raw-svm", "raw-dt", "rule-based", "sax-hac-1nn" and "sax-hac-dt", are used here for comparison.

The "raw-svm" and "raw-dt" methods learn SVM and DT models from the initial 57 labeled instances without any self-learning. The "rule-based" approach is the default mechanism used in current commercial core router systems. When the rule-based method is utilized to identify health status, a wide range of monitored variables of current system are first compared with all existing expert rules, where each rule takes the form "IF {variable $m_i = v$ at time j}, THEN {anomaly $a_p = w_j$}". The number of matched rules and their corresponding severity levels are then determined. Finally, the system is classified into different health levels based on the summarized statistics of all matched rules.

The "sax-hac-1nn" and "sax-hac-dt" are two symbol-based health analysis approaches that are described in [14]. They first transform high-dimensional raw data into low-dimensional symbol sequence. Useful patterns are then learned from these symbol sequences, and 1-nearest-neighbor (1NN) and DT model are utilized to classify various health levels. Figures 5.10 and 5.11 shows the precision and recall for the six health classes using the seven different health-analysis approaches. The results can be summarized as follows:

1. The "self-svm" and "self-dt" methods achieve higher precision and recall than the "raw-svm" and "raw-dt" baselines for all classes. This is because "raw-svm" and "raw-dt" use a limited amount of labeled data to train their models

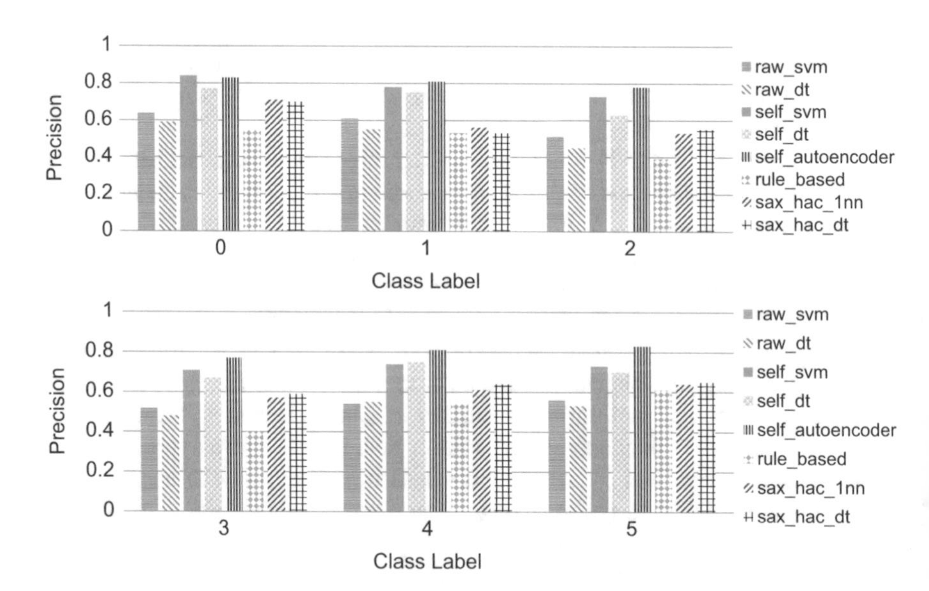

Fig. 5.10 Precision of each class using different methods

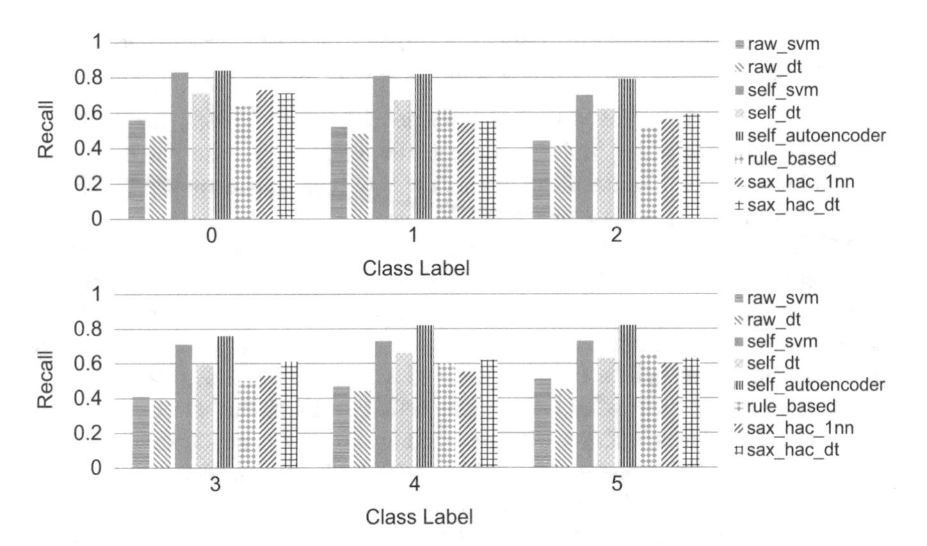

Fig. 5.11 Recall of each class using different methods

while "self-svm" and "self-dt" utilize both labeled and unlabeled data via the self-learning procedure. The "self-svm" method also performs better than the "rule-based" approach because the learned model can identify a wide range of health conditions that are not covered by the existing rule tables.

2. The "sax-hac-1nn" and "sax-hac-dt" approaches achieve better performance than the "raw-svm" and "raw-dt" baselines, but worse performance than the proposed "self-svm" and "self-dt" methods. This is because although a symbol-based approach can reduce high-dimensional feature space without losing critical characteristics, it still suffers from an insufficient amount of labeled data for multi-label classification problems.
3. The two proposed methods, "self-svm" and "self-dt", perform better in classifying "Class 0", "Class 1", "Class 4", "Class 5" than in classifying "Class 2", "Class 3". The reason is that "Class 2" and "Class 3" represent cases where the system is in suspect or warning health conditions. The decision boundaries are thus relatively vague and the learned models are more likely to make incorrect classification. In contrast, "Class 0" and "Class 1" represent the healthy state while "Class 4" and "Class 5" represent severe system illness.
4. The proposed "self-autoencoder" approach provides good classification precision and recall across all health levels. In contrast to the other two proposed methods ("self-svm" and "self-dt"), the "self-autoencoder" does not have any noticeable performance drop in classifying health level 2 and 3. This is because although these two health levels have highly similar time-series characteristics, some critical hidden patterns have been captured from "new" features learned by the LSTM-based autoencoder. Therefore, the "self-autoencoder" can always make

relatively accurate health analysis decision, no matter whether the core routers under test are in healthy, suspect warning or severely ill status.

In summary, when only a limited amount of data are labeled, the proposed feature-based self-learning health analyzer achieves higher precision and recall than the traditional supervised health analyzer as well the currently deployed rule-based health analyzer.

Moreover, root cause analysis of unhealthy core routers can be conducted by interpreting the outputs of the "self-dt" method. Figure 5.12 shows an example of how root cause analysis can be achieved in the proposed "self-dt" method. We can see that after the DT model has been built by our self-learning procedure, a binary tree that consists of both decision nodes and leaf nodes is formed. When a new case i is fed to this "self-dt" model, it starts walking through this tree from the root node until it reaches a leaf node. Its predicted health level is contained in this leaf node, and all the other nodes along this decision path are considered as potential root causes for making this prediction. Therefore, a set of rules is generated from the decision path of the "self-dt" model to provide root-cause suggestion. Since key information such as the name of the component, monitored items (variables) and features have been highlighted in each rule, the expert team can utilize them to effectively locate suspect software/hardware functional units and do further fine-grained root cause analysis by examining the raw time-series. For example, in Fig. 5.12, a set of components such as the block memory of main processing unit have been reported as the root cause candidates for the suspect local behavior of case i.

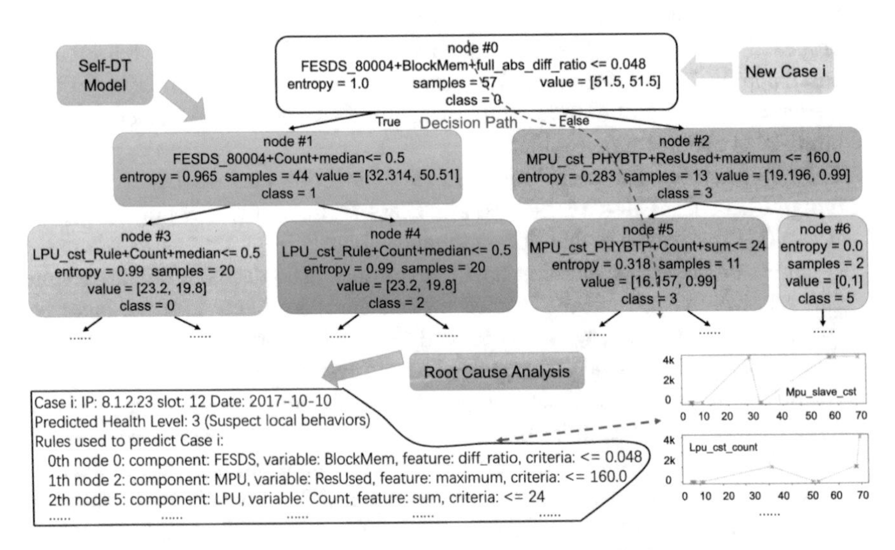

Fig. 5.12 An example of root-cause analysis using the self-DT method

5.5.2 Comparison with Anomaly Detection

Since anomaly detector is the state-of-the-art method used to evaluate the health status of core router systems [15–17], comparing with other anomaly detection approaches is needed to evaluate the effectiveness of the proposed method. However, the metrics used to evaluate the results of anomaly detection and health analysis are considerably different. Therefore, metric normalization was implemented to achieve a fair comparison. Specifically, since anomaly detection can be considered as a binary classification problem while health analysis is a multi-label classification problem, a threshold was introduced into the health analysis procedure to transform a set of health levels $H = \{h_0, h_1, \ldots, h_5\}$ to two-class normal/abnormal labels $A = \{a_0, a_1\}$. Since the number and severity of anomalous behaviors increase from health level h_0 to h_5, a categorical threshold $\{T = j, j = 1, 2, \ldots, 5\}$ is defined to divide H into two groups: $a_0 = \{h_0, \ldots, h_{j-1}\}$ and $a_1 = \{h_j, \ldots, h_5\}$, where a_0 represents the overall normal label and a_1 represents the overall abnormal label. In this way, we can compare anomaly detection baselines with the proposed self-learning health analyzer under the unified binary classification scenario. Moreover, the sensitivity to local or global anomalies can be incorporated by tuning the threshold T. For example, if we take $T = 1$, only the health level h_0 is included in the normal label group a_0 and all the remaining health levels are considered as being anomalous. In this case, both the anomaly detector and health analyzer consider all kinds of anomalies as a whole. In contrast, if $T = 5$ is used, all the health levels except h_5 are considered as normal labels. In this case, both the anomaly detector and health analyzer focus only on identifying severe global anomalous behaviors.

Therefore, in our experiment, we compared the performance of the proposed health analyzer with three anomaly detectors under different threshold values; the results are shown in Figs. 5.13 and 5.14. Four anomaly detection methods, namely "anomaly-knn", "anomaly-hybrid", "anomaly-cp" and "anomaly-autoencoder", and

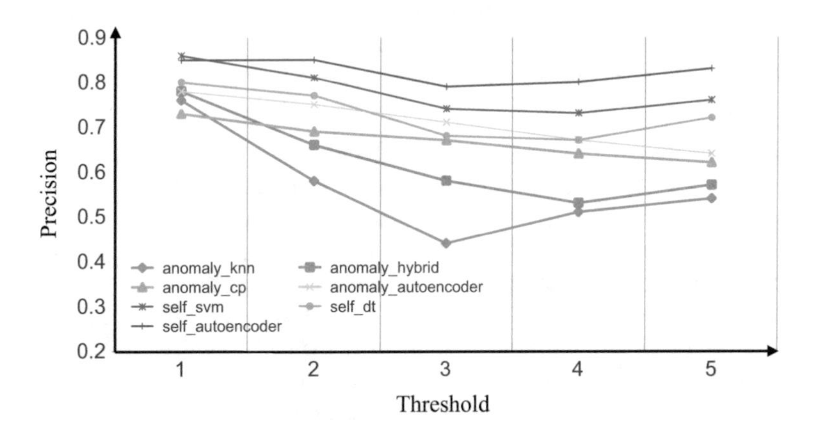

Fig. 5.13 Comparison of precision with anomaly detection methods

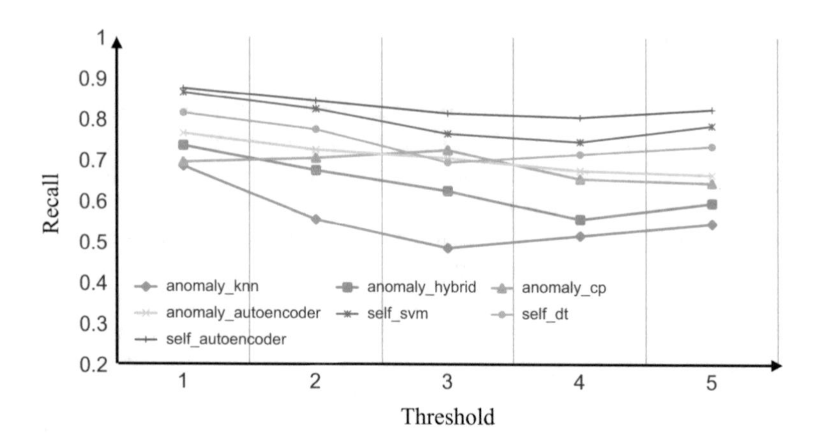

Fig. 5.14 Comparison of recall with anomaly detection methods

three self-learning approaches, namely "self-svm", "self-dt" and "self-autoencoder" are compared in term of both precision and recall. Specifically, "anomaly-knn" is the traditional distance-based anomaly detection method where the distance between the test case and its kth nearest case in the training dataset is calculated as the anomaly score [18]. If the obtained distance exceeds a threshold, this test case is identified as being anomalous. The "anomaly-hybrid" method is a recently proposed feature-categorization-based anomaly detection approach [17]. It utilizes a weighted-voting mechanism to combine results of three kinds of anomaly detectors: distance-based, window-based, and prediction-based so that features are assigned to the fittest anomaly detectors based on their categories and statistical properties. However, since the statistical characteristics of features can change significantly even if no anomalies occur, a changepoint-based anomaly detection method called "anomaly-cp" was proposed in[19]. Various types of changepoints are first detected and a wide range of normal/abnormal patterns are then learned from changepoint windows to help identify local anomalies. The "anomaly-autoencoder" is a recently proposed outlier detection with autoencoder ensembles [20]. The results can be summarized as follows:

1. The "anomaly-knn" method achieves 0.76 precision and 0.69 recall when the threshold $T = 1$. However, its performance drops significantly as T increases to 3. One possible explanation is that as T increases, more cases with suspect health levels are included in the normal label group a_0, making it difficult to distinguish between normal and abnormal groups using distance metrics. The precision and recall recover a little when T increases from 3 to 5. This is because severely unhealthy status exhibits more dissimilarities than other health levels.
2. Although the performance of the "anomaly-hybrid" approach also degrades as T increases, it provides higher accuracy than the "anomaly-knn" method. The reason is that the "anomaly-hybrid" method can automatically assign other types

of anomaly detectors for identifying suspect patterns that cannot be distinguished by distance metrics.
3. In contrast to the above two anomaly detection methods, the "anomaly-cp" approach maintains relatively constant performance as T increases. Its recall even improves as T increases from 1 to 3. One possible reason is that suspect behaviors are usually hidden around changepoints. Therefore, detecting changepoints and learning anomalous behaviors around them is especially effective in identifying local anomalies that exist in both normal and abnormal label groups.
4. The "anomaly-autoencoder" method also maintains relatively good performance as T increases. Moreover, it even achieves similar precision and recall with the "self-dt" approach. This is because a wider range of hidden characteristics can be captured by the ensemble of various autoencoders, making it suitable for detecting hidden local anomalies. However, it performs worse than the "self-svm" and "self-autoencoder" methods due to the partially labeled training instances.
5. Both the "self-svm" and "self-dt" approaches perform better than the first three anomaly detection methods for different values of the threshold T. Moreover, their precision and recall maintain relatively constant as T increases. This is because the proposed health analyzer focuses on fine-grained identification of various health levels, which implicitly covers a wide range of anomalous or suspect behaviors. Meanwhile, the self-learning framework enables effective learning from ambiguous and insufficient labeled data.
6. The "self-autoencoder" method performs even better than the "self-svm" and "self-dt" approaches and its performance is more stable across different thresholds. This is because this method combines the LSTM autoencoder's capability of feature learning and the self-learning framework's capability of label enrichment.

In summary, the two self-learning health analyzers achieve better performance than the three anomaly detection baselines under the transformed binary classification scenario. Moreover, they are sensitive to both local suspect behaviors and global anomalous behaviors, enabling them to remain effective for different threshold values.

5.6 Conclusion

We have presented a feature-based self-learning health analyzer for a complex core router system. First, both the statistical-modeling-based feature extraction and autoencoder-based feature learning have been utilized to capture different characteristics of time-series data. Next, in the self-learning framework, the model for health analysis is iteratively updated using both labeled and unlabeled data. The effectiveness of the health analyzer has been validated using a comprehensive set of field data collected from a set of commercial core routers.

References

1. M. Christ et al., Distributed and parallel time series feature extraction for industrial big data applications (2016). arXiv:1610.07717v3
2. M.V. Wickerhauser, *Adapted Wavelet Analysis from Theory to Software* (IEEE Press, New York, 1994)
3. H. Peng, F. Long, C. Ding, Feature selection based on mutual information: criteria of max-dependency, max-relevance, and min-redundancy. IEEE Trans. Pattern Anal. Mach. Intell. **27**, 1226–1238 (2005)
4. B. Prieto, J.D.L. Asiaín, D. Maravall, Reconfigurable hardware implementation of neural networks for humanoid locomotion, in *International Work-Conference on the Interplay Between Natural and Artificial Computation* (Springer, Berlin, 2005), pp. 395–404
5. F. Ye, Z. Zhang, K. Chakrabarty, X. Gu, Board-level functional fault diagnosis using artificial neural networks, support-vector machines, and weighted-majority voting. IEEE Trans. Comput. Aided Des. Integr. Circuits Syst. **32**, 723–736 (2013)
6. C.-Y. Liou, W.-C. Cheng, J.-W. Liou, D.-R. Liou, Autoencoder for words. Neurocomputing **139**, 84–96 (2014)
7. C.C. Aggarwal, Outlier analysis, in *Data Mining* (Springer, Berlin, 2015), pp. 237–263
8. A. Graves et al., Speech recognition with deep recurrent neural networks, in *International Conference on Acoustics, Speech and Signal Processing* (2013), pp. 6645–6649
9. S. Hochreiter, J. Schmidhuber, Long short-term memory. Neural Comput. **9**, 1735–1780 (1997)
10. S. Miyamoto, A. Terami, Semi-supervised agglomerative hierarchical clustering algorithms with pairwise constraints, in *IEEE International Conference on Fuzzy Systems* (2010), pp. 1–6
11. O. Maimon, L. Rokach, *Data Mining and Knowledge Discovery Handbook* (Springer, Berlin, 2005)
12. R. Giladi, *Network Processors: Architecture, Programming, and Implementation* (Morgan Kaufmann, Los Altos, 2008)
13. V. Antonenko, R. Smelyanskiy, Global network modelling based on mininet approach, in *Proceedings of the Second ACM SIGCOMM Workshop on Hot Topics in Software Defined Networking* (2013), pp. 145–146
14. S. Jin, Z. Zhang, K. Chakrabarty, X. Gu, Symbol-based health-status analysis in a core router system, in *Proceedings of IEEE International Test Conference (ITC)* (2017)
15. A. Patcha, J.-M. Park, An overview of anomaly detection techniques: existing solutions and latest technological trends. Comput. Netw. **51**, 3448–3470 (2007)
16. A. Lazarevic et al., A comparative study of anomaly detection schemes in network intrusion detection, in *Proceedings of the 2003 SIAM International Conference on Data Mining* (2003), pp. 25–36
17. S. Jin et al., Accurate anomaly detection using correlation-based time-series analysis in a core router system, in *Proceedings of IEEE International Test Conference (ITC)* (2016)
18. Y. Liao, V.R. Vemuri, Use of k-nearest neighbor classifier for intrusion detection. Comput. Secur. **21**, 439–448 (2002)
19. S. Jin, Z. Zhang, K. Chakrabarty, X. Gu, Changepoint-based anomaly detection in a core router system, in *Proceedings of IEEE International Test Conference (ITC)* (2017)
20. J. Chen, S. Sathe, C. Aggarwal, D. Turaga, Outlier detection with autoencoder ensembles, in *Proceedings of the 2017 SIAM International Conference on Data Mining* (2017), pp. 90–98

Chapter 6
Conclusions

Over the past decade, advances in both hardware and software techniques have contributed to sustained development and numerous upgrades of core routers that are deployed as backbones in modern IP networks. Core routers nowadays can offer Tbps-level line-rate throughput, store millions of IPV4/6 routes, support hundreds of interfaces, forward billions of packets per second, and consume thousands of Watts power. However, ever-increasing complexity makes core routers more vulnerable to different kinds of failures. It is therefore more difficult to detect and diagnose errors, and more expensive to repair faults. Therefore, a distributed agent-based platform has been designed to monitor the health status of clusters of core routers in IP networks. Anomaly detection and health analysis are the key components needed to achieve such an efficient health monitoring system. This book has covered an array of research related to data-driven anomaly detection and health status analysis in commercial core router systems. Specifically, this book has tackled the problem of how to utilize machine learning and statistical techniques to effectively assess the overall health and identify different types of anomalous behaviors in modern core router systems.

6.1 Book Summary

In this book, we have presented a set of data-driven resiliency solutions for core router systems. Contributions include anomaly detection using correlation-based time series analysis, changepoint-based anomaly detection, hierarchical symbol-based health-status analysis, and self-learning health-status analysis. In contrast to previous anomaly detection methods that relies on expert rules or oversimplified machine learning and statistical techniques, the proposed data-driven methods not only overcome the difficulties in anomaly detection that features have significantly different statistical characteristics and their properties may change as time proceeds,

S. Jin et al., *Anomaly-Detection and Health-Analysis Techniques for Core Router Systems*, https://doi.org/10.1007/978-3-030-33664-6_6

but also provide solutions for analyzing comprehensive health status of core routers from partially-labeled long-term multivariate time-series dataset. The proposed techniques have also been validated using experiments on commercial core routers.

Chapter 2 described an anomaly detector for core router systems using correlation-based time series analysis. We have compared different anomaly detection techniques in terms of their effectiveness for detecting different types of anomalies. We have developed a feature-categorizing-based hybrid method to overcome the difficulty of detecting anomalies in features with different statistical characteristics. We have also implemented a correlation analyzer to remove irrelevant and redundant features as well as identify various types of correlations among extracted features. Real data collected from core router systems have been used to validate the proposed anomaly detector.

Chapter 3 presented a changepoint-based anomaly detector to detect anomalous behaviors even when the statistical properties of the monitored data change significantly as time proceeds. First, we have implemented different changepoint detection approaches to detect different types of changepoints. Next, we have developed a DBSCAN-based clustering method to identify a wide range of normal/abnormal patterns from changepoint windows. Experimental results show that the changepoint-based anomaly detector has achieved better performance than traditional methods.

Chapter 4 introduced a symbol-based health status analyzer to obtain a full picture of the health status of monitored core routers. First, we have implemented multiple symbolization techniques to encode long-term complex time series in a hierarchical way. Next, we have developed several symbol-based clustering and classification methods to identify the health status of core routers. Experimental results show that the proposed symbol-based method has maintained its effectiveness of identifying health status of core routers as the length of time series increases.

Chapter 5 focused on the development of self-learning health analyzer for partially-labeled data extracted from core router systems. First, we have computed a representative feature matrix to capture different characteristics of time-series data. Next, we have utilized hierarchical clustering to infer labels for the unlabeled dataset. Finally, a classifier has been built and iteratively updated using both labeled and unlabeled dataset. The effectiveness of the proposed method has been validated using a comprehensive set of partially-labeled field data collected from a set of commercial core routers.

6.2 Future Directions

Data analytics and real-time monitoring are promising in ensuring that boards and systems operate as intended. Machine-learning and statistical techniques lay the foundation for closing the gap between working silicon and a working system. Inefficient and labor-intensive health diagnosis of high-performance complex system based on expert rules or even manual debug will be replaced by automatic data-driven resilient solutions. This book opens up a number of new research directions related to the prognostic fault tolerance of core router systems as well as other high-performance complex systems. Some potential new directions are summarized below.

6.2.1 Service-Level Data Analysis

As shown in Fig. 1.1, a modern IP network system consists of different types of subnets, where network devices deployed in each subnet can not only have different hardware/software version and configuration, but they can also be linked to each other in various topologies. Moreover, since different subnets may focus on different services, scenario tasks or workloads running on their network devices could also be significantly different. Therefore, service-level (or network-level) data need to be extracted and analyzed so that further health assessment, fault diagnosis, failure prediction, and error recovery for network devices can be made adaptive to different services of networks.

Figure 6.1 shows how service-level data can be extracted from monitored IP network systems. We can see that there are four types of subnets in the figure: carrier ethernet, enterprise network, access network, and transport network. Core routers deployed in these subnets not only have different capabilities and configurations, but also connected with each other in different ways. Since all core routers are monitored by the proposed distributed agent-based monitoring system, an overall status table for core routers can be designed, maintained, and updated in data center. The table mainly consists of three column features: Status, Group, and Description. The Status indicates the overall status of each core router, the Group shows which subnet each core router is deployed at, and the Description includes basic information of each core router, including its device id, running scenario tasks or workloads, etc. These three features can not only help us identify suspect networks and devices, but also provide service-level scenario tasks and workloads information for further health assessment.

6.2.2 Multi-Level Fault Analysis

Fault tolerance for boards and systems can be achieved in two ways: the first is data-driven functional fault diagnosis for boards and the second is data-driven anomaly detection and health assessment for systems. When misbehavior is observed in a running system, fault diagnosis tries to identify the fault that caused it, while anomaly detection tries to identify and assess misbehaviors that may result in a future failure. However, since these two solutions are based on different levels of data, i.e., board-level data and system-level data, respectively, their results as well as methods cannot be directly combined to generate a more comprehensive health

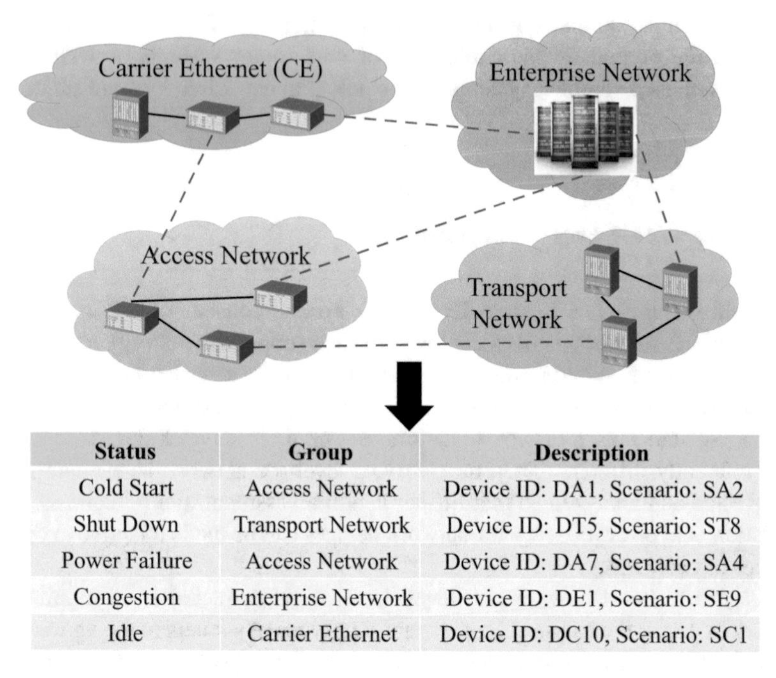

Status	Group	Description
Cold Start	Access Network	Device ID: DA1, Scenario: SA2
Shut Down	Transport Network	Device ID: DT5, Scenario: ST8
Power Failure	Access Network	Device ID: DA7, Scenario: SA4
Congestion	Enterprise Network	Device ID: DE1, Scenario: SE9
Idle	Carrier Ethernet	Device ID: DC10, Scenario: SC1

Fig. 6.1 An example of service-level data extracted from core router clusters

assessment and fault diagnosis in core router systems. Moreover, results obtained from service-level data analysis can also provide useful information for root-cause analysis. Therefore, multi-level fault analysis is needed to identify potential failures in modern IP network systems.

Figure 6.2 shows an example of multi-level fault analysis in modern IP network systems. First, suspect network clusters and links that causes network outages are located by service-level data. Suspect network devices and functional components are then identified by system-level metrics in high-performance computing systems. Finally, suspect SoC and functional units are determined by board-level data from high-speed complex boards. Such a process is time-consuming and labor-intensive because it not only requires analysis of different-level data, from high-level system logs to low-level multimeter records, but also needs cooperation among different operations and functions. For example, Fig. 6.3 presents a wide range of data and operations needed for an automated data-driven multi-level fault analysis engine. We can see that to support a set of core health monitoring and fault analysis functions, data ranging from design process to post-silicon debug are collected while operations covering both forward and backward predictions are implemented.

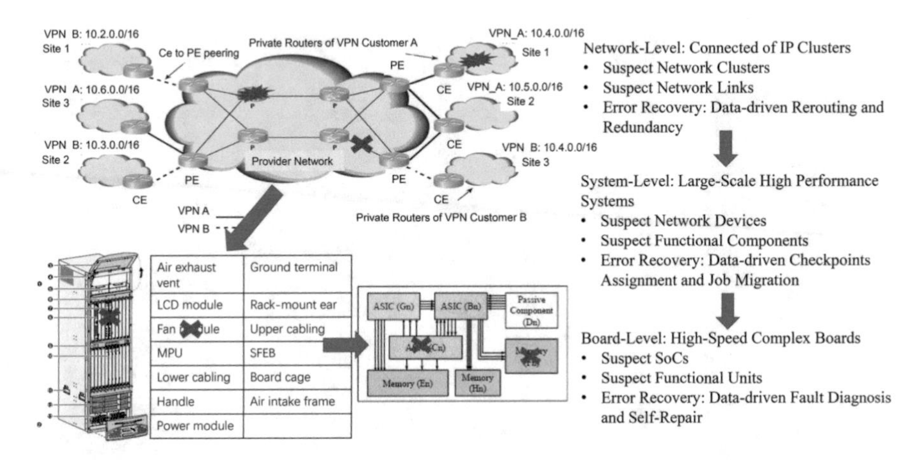

Fig. 6.2 An example of multi-level fault analysis in modern IP network systems

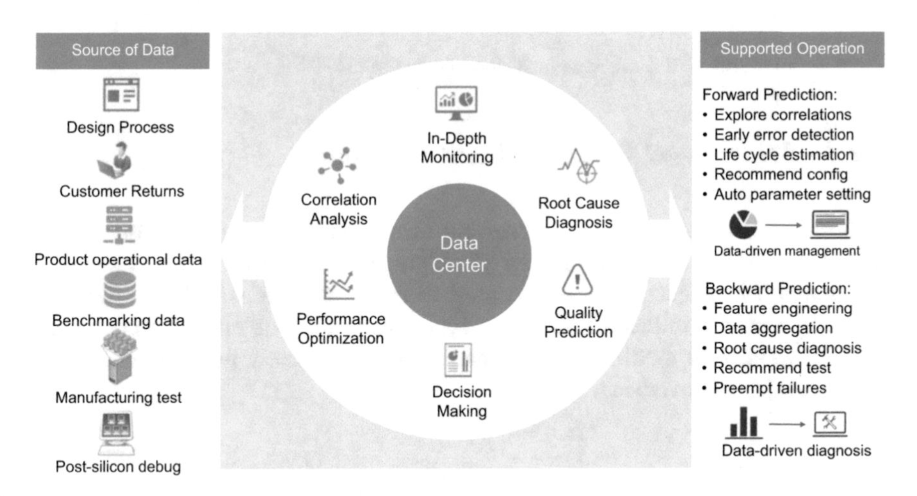

Fig. 6.3 Data source and supported operations in multi-level fault analysis

A more detailed overview of data-driven multi-level fault analysis is shown in Fig. 6.4. In the data pool, data collected from design phase (e.g. design specification, simulation results, etc.), operation phase (e.g. performance monitor, benchmark workload, etc.) and test phase (e.g. test report, repair logs, etc.) are combined in a meaningful way to generate a more informative representation that can be used for backward and forward prediction. The objective of backward prediction is to localize potential faults, give repair suggestions, and take preemptive actions if necessary. In contrast, forward prediction focuses on performance optimization, customer recommendation, and life-cycle estimation.

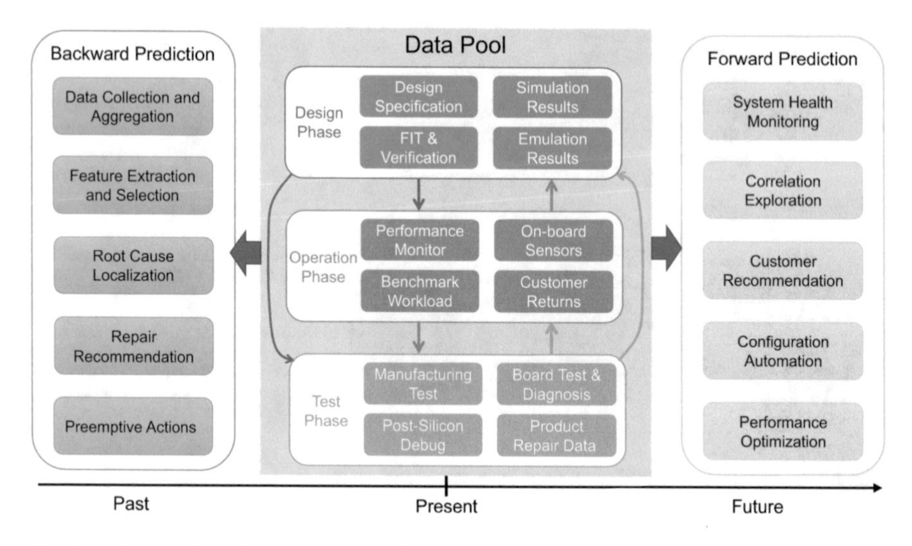

Fig. 6.4 An overview of data-driven multi-level fault analysis

6.2.3 *Multi-Level Proactive Fault Tolerance*

Multi-level fault-tolerant mechanisms are also needed so that a complete data-driven multi-level resilient solution for high-performance complex system can be accomplished. Fault tolerance is used to repair systems with high efficiency, ensuring non-stop utilization. Both the root cause candidate and service scenario interpretation from multi-level fault analyzer can facilitate a more effective error recovery with low overhead.

6.2.3.1 Prediction-Based Checkpoint-Restart Method

In the traditional checkpoint-restart approach, a number of checkpointing nodes are distributed over the execution path of an application. Whenever the application reaches a checkpoint, it will save its current state into that checkpointing node. When a failure occurs, the application restarts from the states stored in its last checkpointing node. A prediction-based checkpoint-restart mechanism can be designed so that new check points will be inserted only if failures are predicted to occur in the near future. Figure 6.5 is a simple illustration of the prediction-based checkpoint-restart mechanism. Node C represents the original checkpoint in the system while node N represents the newly inserted checkpoint. Whenever a decision node D predict the occurrence of failure F in the near future, new node N will be created to save all of the current states. After the actual occurrence of failure F, the application will roll back from F to N instead of from F to C, thereby reducing the overhead of restarting computation.

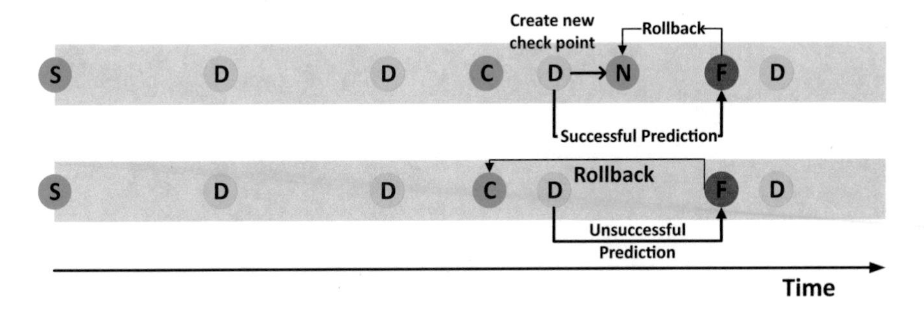

Fig. 6.5 The prediction-based checkpoint-restart approach

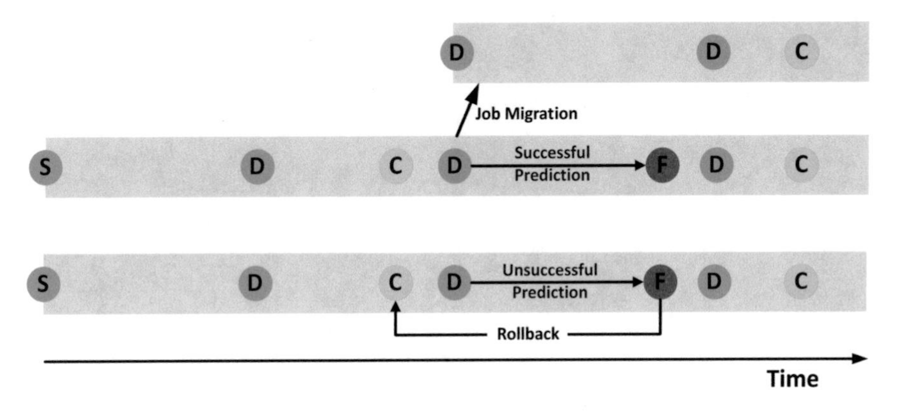

Fig. 6.6 The prediction-based hybrid job-migration method

6.2.3.2 Hybrid Job-Migration Method

In a traditional job-migration method, a number of decision nodes are distributed over the execution path of an application. Whenever a decision node is reached by the application, it will predict whether failures will occur in the near future via analyzing historical system event data. If a failure is predicted to occur, the entire application will be migrated to another fault-free computing component to continue its execution. A hybrid method that combines checkpointing methods and the original job migration approach can be designed to address unpredictable failures. As shown in Fig. 6.6, additional checkpoints are distributed over the execution path. Whenever unpredicted failures occur, the application will restart from last checkpoints instead of stopping execution.

Index

S. Jin et al., *Anomaly-Detection and Health-Analysis Techniques for Core Router Systems*, https://doi.org/10.1007/978-3-030-33664-6

Printed in the United States
By Bookmasters